VERTICAL FARMING

Edited by
Hazelle S. Cabugao

ArclerPress

Vertical Farming

Edited by Hazelle S. Cabugao

Edition 2018

Published by
Arcler Press LLC
708 3rd Avenue, 6th Floor
New York
NY 10017
United States of America
www.arclerpress.com

ISBN: 978-1-68094-566-9 (Hardback)

Printed in United States of America

Library of Congress Control Number: 2017933136

About the Editor

Hazelle S. Cabugao

Hazelle earned her degree in Agriculture from the University of the Philippines Los Baños in 2014. Her studies focus on ornamental horticulture particularly the use of plants and trees in designing landscapes. A horticulturist by profession, she supervises landscape implementation and maintenance and is also passionate about conserving the indigenous flora

Table of Contents

2 TECHNOLOGIES AND DEVICES IN VERTICAL FARMING

3 HORTICULTURAL BUILDING SYSTEM

4 HANGING GARDEN CULTIVATION

5 INTEGRATED FARMING

6 ROOFTOP FARMING

7 WATER CHILLER

List of Abbreviations

ART	Agroforestry Research Trust
AVF	Association for Vertical Farming
BDRI	Biodynamic Research Institute
CDM	Ceramic Discharge Metal Halide
CMH	Ceramic Metal Halide
CSP	Concentrated solar power
CEA	Controlled-Environment Agriculture
CEA	Controlled-environment agriculture
DC	Direct Current
FACE	Free Air Carbon Enrichment
GHG	Greenhouse Gas
HID	High-Intensity Discharge Lamps
H4D	Horticulture for Development
IOH	Institute of Horticulture
ICM	Integrated Crop Management
IF	Integrated Farming
IFM	Integrated Farm Management
IPM	Integrated Pest Management
IWRM	Integrated Water Resources Management
IAPMO	International Association of Plumbing and Mechanical Officials
IFOAM	International Federation of Organic Agriculture Movements
IOBC	International Organisation of Biological Control
ISO	International Organization for Standardization

LCOE	Levelised Cost of Electricity
MCL	Movement for Compassionate Living
PV	Photovoltaic
VRMs	Voltage regulator modules

Preface

Vertical farming or high-rise farming is a proposed indoor, urban farming technology involving large-scale agricultural production in multi-story buildings. It is an intensive farming strategy which mainly employs advanced techniques such as hydroponics and aeroponics to produce crops like fruits, vegetables and edible mushrooms continuously. Unlike traditional farming in non-tropical areas, indoor farming can produce crops year-round. All-season farming multiplies the productivity of the farmed surface by a factor of 4 to 6 depending on the crop. With some crops, such as strawberries, the factor may be as high as 30. Furthermore, as the crops would be sold in the same infrastructures in which they are grown, they will not need to be transported between production and sale, resulting in less spoilage, infestation, and energy required than conventional farming encounters. Crops grown in traditional outdoor farming suffer from the often suboptimal, and sometimes extreme, nature of geological and meteorological events such as undesirable temperatures or rainfall amounts, monsoons, hailstorms, tornadoes, flooding, wildfires, and severe droughts. The protection of crops from weather is increasingly important as global climate change occurs. Because vertical plant farming provides a controlled environment, the productivity of vertical farms would be mostly independent of weather and protected from extreme weather events. Although the controlled environment of vertical farming negates most of these factors, earthquakes and tornadoes still pose threats to the proposed infrastructure, although this again depends on the location of the vertical farms. With high-rise farming, a relatively large area of land will be converted into a facility on which a multi-story building will be constructed. It will be located in the urban center. Important food crops will be grown in this building on soil-less media, employing mainly the techniques in hydroponics. It is estimated that by the year 2050, close to 80% of the world's population will live in urban areas and the total population of the world will increase by 3 billion people. A very large amount of land may be required depending on the change in yield per hectare. Scientists are concerned that this large amount of required farmland will not be available and that severe damage to the earth will be caused by the added farmland.

Content Coverage

Chapter One is intended to present the introduction of producing food in vertically stacked layers, such as in a skyscraper, used warehouse, or shipping container. The modern ideas of vertical farming use indoor farming techniques and controlled-environment agriculture (CEA) technology, where all environmental factors can be controlled.

Chapter Two highlights about technologies and devices in vertical farming, such as greenhouse is a structure with walls and roof made chiefly of transparent material, in which plants requiring regulated climatic conditions are grown.

Chapter Three examines the horticultural building systems, where vegetation and an architectural system exist in a mutually defined and intentionally designed relationship that supports plant growth and an architectonic concept. The most common form of these systems in contemporary vernacular is green wall, vertical garden, green roof, roof garden, building-integrated agriculture.

Chapter Four highlights about the hanging gardens, which is a sustainable landscape architecture, an artistic garden or a small urban farm: attached to or built on a wall. They mainly found in areas where land is scarce or where the farmer is mobile or not permanent.

Chapter Five reveals the integrated farming system, which enables the farmers to identify opportunities and threats and act accordingly, and, at the same time, consider consumer interests in their business.

Chapter Six emphasizes on the practice of cultivating food on the rooftop of buildings, sometimes referred to as rooftop farming. Rooftop farming is usually done using green roof, hydroponics, aeroponics or air-dynaponics systems or container gardens.

Chapter Seven focuses on water chiller mechanism used to facilitate heat exchange from water to a refrigerant in a closed loop system. In hydroponics, pumps, lights and ambient heat can warm the reservoir water temperatures, leading to plant root and health problems. This results in healthy root production and efficient absorption of nutrients.

CHAPTER 1

Vertical Farming

Chapter Contents

- **Introduction**
- **Types**
- **Problems**
- **Biodynamic Agriculture**
- **Climate-Friendly Gardening**
- **Square Foot Gardening**
- **Vegan Organic Gardening**

Vertical farming or high-rise farming is a proposed indoor, urban farming technology involving large-scale agricultural production in multi-story buildings. It is an intensive farming strategy which mainly employs advanced techniques such as hydroponics and aeroponics to produce crops like fruits, vegetables and edible mushrooms continuously. It also includes the raising of livestock and fishes indoor.

INTRODUCTION

Vertical farming is the practice of producing food in vertically stacked layers, such as in a skyscraper, used warehouse, or shipping container. The modern ideas of vertical farming use indoor farming techniques and controlled-environment agriculture (CEA) technology, where all environmental factors can be controlled. These facilities utilize artificial control of light, environmental control (humidity, temperature, gases...) and fertigation. Some vertical farms use techniques similar to greenhouses, where natural sunlight can be augmented with artificial lighting and metal reflectors.

Lettuce grown in indoor vertical farming system.

TYPES

The term "vertical farming" was coined by Gilbert Ellis Bailey in 1915 in his book *Vertical Farming*. His use of the term differs from the current meaning—he wrote about farming with a special interest in soil origin, its nutrient content and the view of plant life as "vertical" life forms, specifically relating to their underground root structures. Modern usage of the term "vertical farming" usually refers to growing plants in layers, whether in a multistory skyscraper, used warehouse, or shipping container.

Mixed-use Skyscrapers

Mixed-use skyscrapers were proposed and built by architect Ken Yeang. Yeang proposes that instead of hermetically sealed mass-produced agriculture, plant life should be cultivated within open air, mixed-use skyscrapers for climate control and consumption. This version of vertical farming is based upon personal or community use rather than the wholesale production and distribution plant life that aspires to feed an entire city. It thus requires less of an initial investment than Despommier's "vertical farm". However, neither Despommier nor Yeang are the conceptual originators, nor is Yeang the inventor of vertical farming in skyscrapers.

Despommier's Skyscrapers

Dickson Despommier shares his ideas about how "vertical farming" can help reduce hunger by changing the way we use land for agriculture. Photography by Kris Krüg.

Ecologist Dickson Despommier argues that vertical farming is legitimate for environmental reasons. He claims that the cultivation of plant life within skyscrapers will require less embedded energy and produce less pollution than some methods of producing plant life on natural landscapes. He moreover claims that natural landscapes are too toxic for natural, agricultural production, despite the ecological and environmental costs of extracting materials to build skyscrapers for the simple purpose of agricultural production.

Vertical farming according to Despommier thus discounts the value of natural landscape in exchange for the idea of skyscraper as spaceship. Plant life is mass-produced within hermetically sealed, artificial environments that have little to do with the outside world. In this sense, they could be built anywhere regardless of the context. Although climate control, lighting, and other costs of maintenance have been posited as potentially stifling to bringing this concept to fruition, advocates have countered that an important feature of future vertical farms will be the integration of renewable energy technology, be it solar panels, wind turbines, water capture systems, and probably some combination of the three. The vertical farm is designed to be sustainable, and to enable nearby inhabitants to work at the farm.

Despommier's concept of the vertical farm emerged in 1999 at Columbia University. It promotes the mass cultivation of plant life for commercial purposes in skyscrapers.

Stackable Shipping Containers

Several companies have brought forth the concept of stacking recycled shipping containers in urban settings. Freight Farms produces a "leafy green machine" that is a complete farm-to-table system outfitted with vertical hydroponics, LED lighting and intuitive climate controls built within a 12 m × 2.4 m shipping container. Podponics has built a large scale vertical farm in Atlanta consisting of over 100 stacked "growpods". A similar farm is currently under construction in Oman.

A commercial high-rise farm such as the 'Vertical Farm' has never been built, yet extensive photographic documentation and several historical books on the subject suggest that research on the subject was not diligently pursued. New sources indicate that a tower hydroponicum existed in Armenia prior to 1951.

Proponents argue that, by allowing traditional outdoor farms to revert to a natural state and reducing the energy costs needed to transport foods to consumers, vertical farms could significantly alleviate climate change produced by excess atmospheric carbon. Critics have noted that the costs of the additional energy needed for artificial lighting, heating and other vertical farming operations would outweigh the benefit of the building's close proximity to the areas of consumption. However, a recent study published in the Journal of Agricultural Engineering and Biotechnology has utilized inexpensive metal reflectors to supply sunlight to the plants.

One of the earliest drawings of a tall building that cultivates food was published in Life Magazine in 1909. The reproduced drawings feature vertically stacked homesteads set amidst a farming landscape. This proposal can be seen in Rem Koolhaas's Delirious New York. Koolhaas wrote that this 1909 theorem is ‹The Skyscraper as Utopian device for the production of unlimited numbers of virgin sites on a metropolitan location' (1994, 82).

Other architectural proposals that provide the seeds for the Vertical Farm project include Le Corbusier's Immeubles-Villas (1922) and SITE's Highrise of homes (1972). SITE's Highrise of homes, is a near revival of the 1909 Life Magazine Theorem. In fact, built examples of tower hydroponicums are quite well documented in the canonical text of The Glass House by John Hix. Images of the vertical farms at the School of Gardeners in Langenlois, Austria, and the glass tower at the Vienna International Horticulture Exhibition (1964) clearly show that vertical farms existed more than 40 years prior to contemporary discourse on the subject. Although architectural precedents remain valuable, the technological precedents that make vertical farming possible can be traced back to horticultural history through the development of greenhouse and hydroponic technology. Early building types or Hydroponicums were developed, integrating hydroponic technology into building systems. These horticultural building systems evolved from greenhouse technology, and paved the way for the modern concept of the vertical farm. The British Interplanetary Society developed a hydroponicum for lunar conditions and other building prototypes were developed during the early days of space exploration. During this era of expansion and experimentation, the first Tower Hydroponic Units were developed in Armenia.

The Armenian tower hydroponicums are the first built examples of a vertical farm, and is documented in Sholto Douglas' seminal text Hydroponics: The Bengal System first published in 1951 with data from the then-East Pakistan, today's Bangladesh, and the Indian state of West Bengal. Contemporary notions of vertical farming are predated by this early technology by more than 50 years. Contemporary precursors that have been published, or built, are Ken Yeang's Bioclimatic Skyscraper (Menara Mesiniaga, built

1992); MVRDV's PigCity, 2000; MVRDV's Meta City/ Datatown (1998–2000); Pich-Aguilera's Garden Towers (2001).

Ken Yeang is perhaps the most widely known architects that has promoted the idea of the ‹mixed-use' Bioclimatic Skyscraper which combines living units and opportunities for food production.

Early prototypes of vertical farms, or Tower Hydroponicums existed in Armenia prior to 1951 during an era of hydroponic and horticultural building system research fueled by space exploration and a transatlantic technology race.

The latest version of these very idea is Dickson Despommier's The Vertical Farm.

Dickson Despommier, a professor of environmental health sciences and microbiology at Columbia University in New York City, modernized the idea of vertical farming in 1999 with graduate students in a medical ecology class. Although much of Despommier's suggestions have been challenged and strongly criticized from an environmental science and engineering point of view, the idea's popularization in recent years has been largely the result of Despommier's assertion that food production can be transformed.

Despommier had originally challenged his class to feed the entire population of Manhattan (About 2,000,000 people) using only 5 hectares (13 acres) of usable rooftop gardens. The class calculated that, by using rooftop gardening methods, only 2 percent of the population would be fed. Unsatisfied with the results, Despommier made an off-the-cuff suggestion of growing plants indoors, vertically. The idea sparked the students' interests and gained major momentum. By 2001 the first outline of a vertical farm was introduced and today scientists, architects, and investors worldwide are working together to make the concept of vertical farming a reality. In an interview with Miller-McCune.com, Despommier described how vertical farms would function:

Architectural designs have been produced by Chris Jacobs and Andrew Kranis from Columbia University and Gordon Graff from the University of Waterloo's School of Architecture in Cambridge, ON.

Mass media attention began with an article written in *New York* magazine. Since 2007, articles have appeared in *The New York Times*, *U.S. News & World Report*, *Popular Science*, *Scientific American* and *Maxim*, among others, as well as radio and television features. As of 2012, Vertical Harvest is working on raising funds for an urban, small-scale vertical farm in Jackson Hole, Wyoming.

In 2013, July 18, the Association for Vertical Farming was founded in Munich, Germany As of 2014, Vertical Fresh Farms has been operating in Buffalo, NY specializing in a wide variety of salad greens, herbs, and sprouts.

PROBLEMS

Economics

Opponents question the potential profitability of vertical farming. At current levels of technology the ability of vertical farms to compete with established farming processes is limited. The extra cost of lighting, heating, and powering the vertical farm may negate any of the cost benefits received by the decrease in transportation expenses. The economic and environmental benefits of vertical farming rest partly on the concept of minimizing food miles, the distance that food travels from farm to consumer. However, a recent analysis suggests that transportation is only a minor contributor to the economic and environmental costs of supplying food to urban populations. The author of the report, University of Toronto professor Pierre Desrochers, concluded that "food miles are, at best, a marketing fad." Thus the facility would have to produce a considerable profit to justify remaining in the city. A simpler concept rather than trying to stack farms on multiple stories would be to just cultivate crops on the roofs of existing building. Rooftop farming is a growing urban trend, requires little construction (other than fortifying the roof to hold the weight of the growing medium), still takes advantage of sunlight and doesn't require investment in machinery, growing lights or irrigation.

Similarly, if the power needs of the vertical farm are met by fossil fuels, the environmental effect may be a net loss; even building low-carbon capacity to power the farms may not make as much sense as simply leaving the traditional farms in place, and burning less coal.

The initial building costs will be easily over $100 million, for a 60 hectare vertical farm. Office occupancy costs can be very high in major cities, with cities such as Tokyo, Moscow, Mumbai, Dubai, Milan, Zurich, and Sao Paulo ranging from $1850 to $880 per square meter, respectively.

Energy use

During the growing season, the sun shines on a vertical surface at an extreme angle such that much less light is available to crops than when they are planted on flat land. Therefore, supplemental light would be required in order to obtain economically viable yields. Bruce Bugbee, a crop physiologist at Utah State University, believes that the power demands of vertical farming will be too expensive and uncompetitive with traditional farms using only free natural light. The environmental writer George Monbiot calculated that the cost of providing enough supplementary light to grow the grain for a single loaf would be about $15. An article in the Economist argued that "even though crops growing in a glass skyscraper will get some natural sunlight during the day, it won't be enough" and "the cost of powering artificial lights will make indoor farming prohibitively expensive".

As "The Vertical Farm" proposes a controlled environment, heating and cooling costs will be at least as costly as any other tower. But there also remains the issue of complicated, if not more expensive, plumbing and elevator systems to distribute food and water throughout. Even throughout the northern continental United States, while heating with relatively cheap fossil fuels, the heating cost can be over $200,000 per hectare.

To address this problem, The Plant in Chicago is building an anaerobic digester into the building. This will allow the farm to operate off the energy grid. Moreover, the anaerobic digester will be recycling waste from nearby businesses that would otherwise go into landfills.

Pollution

Depending on the method of electricity generation used, regular greenhouse produce can create more greenhouse gases than field produce, largely due to higher energy use per kilogram of produce. With vertical farms requiring much greater energy per kilogram of produce, mainly through increased lighting, than regular greenhouses, the amount of pollution created will be much higher than that from field produce. The amount of pollution produced is dependent on how the energy used in the process is generated.

As plants acquire nearly all their carbon from the atmosphere, greenhouse growers commonly supplement CO_2 levels to 3–4 times the rate normally found in the atmosphere. This increase in CO_2, which has been shown to increase photosynthesis rates by 50%, contributes to the higher yields expected in vertical farming. It is not uncommon to find greenhouses burning fossil fuels purely for this purpose, as other CO_2 sources, like from furnaces, contain pollutants such as sulphur dioxide and ethylene which significantly damage plants. This means a vertical farm will require a CO_2 source, most likely from combustion, even if the rest of the farm is powered by "green" energy. Also, through necessary ventilation, much CO_2 will be leaked into the city's atmosphere.

Greenhouse growers commonly exploit photoperiodism in plants to control whether the plants are in a vegetative or reproductive stage. As part of this control, growers will have the lights on past sunset and before sunrise or periodically throughout the night. Single story greenhouses are already a nuisance to neighbours because of light pollution, a 30-story vertical farm in a densely populated area will surely face problems because of its light pollution.

Hydroponics greenhouses regularly change the water, meaning there is a large quantity of water containing fertilizers and pesticides that must be disposed of. While solutions are currently being worked on, the most common method of simply spreading the mixture over a sufficient area of neighbouring farmland or wetlands would be more difficult for an urban vertical farm.

Advantages

Several potential advantages of vertical farming have been discussed by Despommier. Many of these benefits are obtained from scaling up hydroponic or aeroponic growing methods.

Preparation for the future

It is estimated that by the year 2050, close to 80% of the world's population will live in urban areas and the total population of the world will increase by 3 billion people. A very large amount of land may be required depending on the change in yield per hectare. Scientists are concerned that this large amount of required farmland will not be available and that severe damage to the earth will be caused by the added farmland. According to Despommier, vertical farms, if designed properly, may eliminate the need to create additional farmland and help create a cleaner environment.

Increased crop production

Unlike traditional farming in non-tropical areas, indoor farming can produce crops year-round. All-season farming multiplies the productivity of the farmed surface by a factor of 4 to 6 depending on the crop. With some crops, such as strawberries, the factor may be as high as 30.

Furthermore, as the crops would be sold in the same infrastructures in which they are grown, they will not need to be transported between production and sale, resulting in less spoilage, infestation, and energy required than conventional farming encounters. Research has shown that 30% of harvested crops are wasted due to spoilage and infestation, though this number is much lower in developed nations.

Despommier suggests that, if dwarf versions of certain crops are used (e.g. dwarf wheat, which has been grown in space by NASA, is smaller in size but richer in nutrients), year-round crops, and "stacker" plant holders are accounted for, a 30-story building with a base of a building block (2 hectares (5 acres)) would yield a yearly crop analogous to that of 1,000 hectares (2,400 acres) of traditional farming.

Protection from Weather-Related Problems

Crops grown in traditional outdoor farming suffer from the often suboptimal, and sometimes extreme, nature of geological and meteorological events such as undesirable temperatures or rainfall amounts, monsoons, hailstorms, tornadoes, flooding, wildfires, and severe droughts. The protection of crops from weather is increasingly important as global climate change occurs. "Three recent floods (in 1993, 2007 and 2008) cost the United States billions of dollars in lost crops, with even more devastating losses in topsoil. Changes in rain patterns and temperature could diminish India's agricultural output by 30 percent by the end of the century."

Because vertical plant farming provides a controlled environment, the productivity of vertical farms would be mostly independent of weather and protected from extreme weather events. Although the controlled environment of vertical farming negates most of these factors, earthquakes and tornadoes still pose threats to the proposed infrastructure, although this again depends on the location of the vertical farms.

Conservation of Resources

Each unit of area in a vertical farm could allow up to 20 units of area of outdoor farmland to return to its natural state,and recover farmlands due to development from original flat farmlands.

Vertical farming would reduce the need for new farmland due to overpopulation, thus saving many natural resources,currently threatened by deforestation or pollution. Deforestation and desertification caused by agricultural encroachment on natural biomes would be avoided. Because vertical farming lets crops be grown closer to consumers, it would substantially reduce the amount of fossil fuels currently used to transport and refrigerate farm produce. Producing food indoors reduces or eliminates conventional plowing, planting, and harvesting by farm machinery, also powered by fossil fuels.

Halting Mass Extinction

Withdrawing human activity from large areas of the Earth's land surface may be necessary to slow and eventually halt the current anthropogenic mass extinction of land animals.

Traditional agriculture is highly disruptive to wild animal populations that live in and around farmland and some argue it becomes unethical when there is a viable alternative. One study showed that wood mouse populations dropped from 25 per hectare to 5 per hectare after harvest, estimating 10 animals killed per hectare each year with conventional farming. In comparison, vertical farming would cause very little harm to wildlife, and would allow disused farmland to return to its pre-agricultural state.

Impact on Human Health

Traditional farming is a hazardous occupation with particular risks that often take their toll on the health of human laborers. Such risks include: exposure to infectious diseases such as malaria and schistosomes, exposure to toxic chemicals commonly used as pesticides and fungicides, confrontations with dangerous wildlife such as venomous snakes, and the severe injuries that can occur when using large industrial farming equipment. Whereas the traditional farming environment inevitably contains these risks (particularly in the farming practice known as "slash and burn"), vertical farming – because the environment is strictly controlled and predictable reduces some of these dangers. Currently, the American food system makes fast, unhealthy food cheap while fresh produce is less available and more expensive, encouraging poor eating habits. These poor eating habits lead to health problems such as obesity, heart disease, and diabetes. The increased availability and subsequent lower cost of fresh produce would encourage healthier eating.

Poverty/Destitution and Culture

Food security is one of the primary factors leading to absolute poverty. Being able to construct 'farm land' in secure areas as needed will help alleviate the pressures causing crises among neighbors fighting for resources (mainly water and space). It also allows continued growth of culturally significant food items without sacrificing sustainability or basic needs, which can be significant to the recovery of a society from poverty.

Urban Growth

Vertical farming, used in conjunction with other technologies and socioeconomic practices, could allow cities to expand while remaining largely self-sufficient food wise. This would allow for large urban centers that could grow without destroying considerably larger areas of forest to provide food for their people. Moreover, the industry of vertical farming will provide employment to these expanding urban centers. This may help displace the unemployment created by the dismantling of traditional farms, as more farm laborers move to cities in search of work.

Energy Sustainability

Vertical farms could exploit methane digesters to generate a small portion of its own electrical needs. Methane digesters could be built on site to transform the organic waste generated at the farm into biogas which is generally composed of 65% methane along with other gases. This biogas could then be burned to generate electricity for the greenhouse.

Technologies and Devices

Vertical farming relies on the use of various physical methods to become effective. Combining these technologies and devices in an integrated whole is necessary to make Vertical Farming a reality. Various methods are proposed and under research. The most common technologies suggested are:

- Greenhouse
- The Folkewall and other vertical growing architectures
- Flowerpot
- Aeroponics / Hydroponics / Aquaponics
- Composting
- Grow light
- Phytoremediation
- Skyscraper
- Controlled-environment agriculture
- Precision agriculture
- Agricultural robot

Plans

Despommier argues that the technology to construct vertical farms currently exists. He also states that the system can be profitable and effective, a claim evidenced by some preliminary research posted on the project's website. Developers and local governments in the following cities have expressed serious interest in establishing a vertical farm: Incheon (South Korea), Abu Dhabi (United Arab Emirates), and Dongtan (China), New York City, Portland, Ore., Los Angeles, Las Vegas, Seattle, Surrey, B.C., Toronto, Paris, Bangalore, Dubai, Shanghai and Beijing. The Illinois Institute of Technology is now crafting a detailed plan for Chicago. It is suggested that prototype versions of vertical farms should be created first, possibly at large universities interested in the research of vertical farms, in order to prevent failures such as the Biosphere 2 project in Oracle, Arizona.

In 2009, the world's first pilot production system was installed at Paignton Zoo Environmental Park in the United Kingdom. The project showcased a technological solution for vertical farming and provided a physical base to conduct research into sustainable urban food production. The produce is used to feed the zoo's animals while the project enables evaluation of the systems and provides an educational resource to advocate for change in unsustainable land use practices that impact upon global biodiversity and ecosystem services,

In 2010, the Green Zionist Alliance proposed a resolution at the 36th World Zionist Congress calling on Keren Kayemet L'Yisrael (Jewish National Fund in Israel) to develop vertical farms in Israel.

In 2012, the world's first commercial vertical farm was opened in Singapore, developed by Sky Greens Farms, and is three stories high. They currently have over 100 towers that stand at nine meters tall.

In 2013 (July 18) the Association for Vertical Farming (AVF) was founded in Munich (Germany). By May 2015 the AVF already expanded with regional chapters all over Europe, Asia, USA, Canada and the United Kingdom. This internationally active non-profit organization unites growers and inventors to improve food security and the sustainable development of Vertical Farming. The AVF also focuses on advancing vertical farming technologies, designs and businesses by hosting international info-days, workshops and summits. They developed a glossary to bring consistency to the industry and plan on helping to standardize the technologies.

BIODYNAMIC AGRICULTURE

Biodynamic agriculture is a form of alternative agriculture very similar to organic farming, but which includes various esoteric concepts drawn from the ideas of Rudolf Steiner (1861–1925). Initially developed in the 1920s, it was the first of the organic agriculture movements. It treats soil fertility, plant growth, and livestock care as ecologically interrelated tasks, emphasizing spiritual and mystical perspectives.

Biodynamics has much in common with other organic approaches – it emphasizes the use of manures and composts and excludes the use of artificial chemicals on soil and plants. Methods unique to the biodynamic approach include its treatment of animals, crops, and soil as a single system; an emphasis from its beginnings on local production and distribution systems; its use of traditional and development of new local breeds and varieties; and the use of an astrological sowing and planting calendar. Biodynamic agriculture uses various herbal and mineral additives for compost additives and field sprays; these are sometimes prepared by controversial methods, such as burying ground quartz stuffed into the horn of a cow, which are said to harvest cosmic forces in the soil, that are more akin to sympathetic magic than agronomy.

As of 2016 biodynamic techniques were used on 161,074 hectares in 60 countries. Germany accounts for 45% of the global total; the remainder average 1750 ha per country. Biodynamic methods of cultivating grapevines have been taken up by several notable vineyards. There are certification agencies for biodynamic products, most of which are members of the international biodynamics standards group Demeter International.

No difference in beneficial outcomes has been scientifically established between certified biodynamic agricultural techniques and similar organic and integrated farming practices. Critics have characterized biodynamic agriculture as pseudoscience on the basis of a lack of strong evidence for its efficacy and skepticism about aspects characterized as magical thinking.

Biodynamics was the first modern organic agriculture. Its development began in 1924 with a series of eight lectures on agriculture given by philosopher Rudolf Steiner at Schloss Koberwitz in Silesia, Germany, (now Kobierzyce in Poland southwest of Wrocław). These lectures, the first known presentation of organic agriculture, were held in response to a request by farmers who noticed degraded soil conditions and a deterioration in the health and quality of crops and livestock resulting from the use of chemical fertilizers. The one hundred eleven attendees, less than half of whom were farmers, came from six countries, primarily Germany and Poland. The lectures were published in November 1924; the first English translation appeared in 1928 as *The Agriculture Course*.

Steiner emphasized that the methods he proposed should be tested experimentally. For this purpose, Steiner established a research group, the "Agricultural Experimental Circle of Anthroposophical Farmers and Gardeners of the General Anthroposophical Society". This research group attracted, in the interval 1924 to 1939, about 800 members from around the world, including Europe, the Americas and Australasia.Another group, the "Association for Research in Anthroposophical Agriculture" (Versuchsring anthroposophischer Landwirte), directed by the German agronomist Erhard Bartsch, was formed to test the effects of biodynamic methods on the life and health of soil, plants and animals; the group published a monthly journal *Demeter*.Bartsch was also instrumental in developing a sales organisation for biodynamic products, Demeter, which still exists today. The Research Association was renamed The Imperial Association for Biodynamic Agriculture (Reichsverband für biologisch-dynamische Wirtschaftsweise) in 1933. It was dissolved by the National Socialist regime in 1941. In 1931 the association had 250 members in Germany, 109 in Switzerland, 104 in other European countries and 24 outside Europe. The oldest biodynamic farms are the Wurzerhof in Austria and Marienhöhe in Germany.

In 1938, Ehrenfried Pfeiffer's text *Bio-Dynamic Farming and Gardening* was published in five languages – English, Dutch, Italian, French, and German; this became the standard work in the field for several decades. In July 1939, at the invitation of Walter James, 4th Baron Northbourne, Pfeiffer travelled to the UK and presented the Betteshanger Summer School and Conference on Biodynamic Farming' at Northbourne's farm in Kent. The conference has been described as the ‹missing link' between biodynamic agriculture and organic farming because, in the year after Betteshanger, Northbourne published his manifesto of organic farming, *Look to the Land*, in which he coined the term 'organic farming' and praised the methods of Rudolf Steiner. In the 1950s, Hans Mueller was encouraged by Steiner's work to create the organic-biological farming method in Switzerland; this later developed to become the largest certifier of organic products in Europe, *Bioland*.[:5]

Today biodynamics is practiced in more than 50 countries worldwide and in a variety of circumstances, ranging from temperate arable farming, viticulture in France, cotton production in Egypt, to silkworm breeding in China.[:141] Germany accounts for nearly half of the world's biodynamic agriculture. Demeter International is the primary certification agency for farms and gardens using the methods.

Geographic Developments

- In Australia, the first biodynamic farmer was Ernesto Genoni who in 1928 joined the Experimental Circle of Anthroposophical Farmers and Gardeners, followed soon after by his brother Emilio Genoni. Bob Williams presented the first public lecture in Australia on biodynamic agriculture on 26 June 1938 at the home of the architects Walter Burley Griffin and Marion Mahony Griffin at Castlecrag, Sydney. Since the 1950s research work has continued at the Biodynamic Research Institute (BDRI) in Powelltown, near Melbourne under the direction of Alex Podolinsky. In 1989 Biodynamic Agriculture Australia was established, as a not for profit association.
- In 1928 the *Anthroposophical Agricultural Foundation* was founded in England; this is now called the *Biodynamic Agriculture Association*. In 1939, Britain's first biodynamic agriculture conference, the Betteshanger Summer School and Conference on Biodynamic Agriculture, was held at Lord Northbourne's farm in Kent; Ehrenfried Pfeiffer was the lead presenter.
- In the United States, the Biodynamic Farming & Gardening Association was founded in 1938 as a New York state corporation.
- In France the International Federation of Organic Agriculture Movements (IFOAM) was formed in 1972 with five founding members, one of which was the Swedish Biodynamic Association.
- The University of Kassel had a Department of Biodynamic Agriculture from 2006 to March 2011.

Biodynamic Method of Farming

In common with other forms of organic agriculture, biodynamic agriculture uses management practices that are intended to "restore, maintain and enhance ecological harmony." Central features include crop diversification, the avoidance of chemical soil treatments and off-farm inputs generally, decentralized production and distribution, and the consideration of celestial and terrestrial influences on biological organisms. The Demeter Association recommends that "(a) minimum of ten percent of the total farm acreage be set aside as a biodiversity preserve. That may include but is not limited to forests, wetlands, riparian corridors, and intentionally planted insectaries. Diversity in crop rotation and perennial planting is required: no annual crop can be planted in the same field for more than two years in succession. Bare tillage year round is prohibited so land needs to maintain adequate green cover."

The Demeter Association also recommends that the individual design of the land "by the farmer, as determined by site conditions, is one of the basic tenets of biodynamic agriculture. This principle emphasizes that humans have a responsibility for the development of their ecological and social environment which goes beyond economic aims and the principles of descriptive ecology." Crops, livestock, and farmer, and "the entire socioeconomic environment" form a unique interaction, which biodynamic farming tries to "actively shape ...through a variety of management practices. The prime objective is always to encourage healthy conditions for life": soil fertility, plant and animal health, and product quality.[:141–142]"The farmer seeks to enhance and support the forces of nature that lead to healthy crops, and rejects farm management practices that damage the environment, soil plant, animal or human health....the farm is conceived of as an organism, a self-contained entity with its own individuality,"[:148] holistically conceived and self-sustaining. "Disease and insect control are addressed through botanical species diversity, predator habitat, balanced crop nutrition, and attention to light penetration and airflow. Weed control emphasizes prevention, including timing of planting, mulching, and identifying and avoiding the spread of invasive weed species."

Biodynamic agriculture differs from many forms of organic agriculture in its spiritual, mystical, and astrological orientation. It shares a spiritual focus, as well as its view toward improving humanity, with the nature farming movement in Japan.Important features include the use of livestock manures to sustain plant growth (recycling of nutrients), maintenance and improvement of soil quality, and the health and well being of crops and animals. Cover crops, green manures and crop rotations are used extensively and the farms to foster the diversity of plant and animal life, and to enhance the biological cycles and the biological activity of the soil.

Biodynamic farms often have a cultural component and encourage local community, both through developing local sales and through on-farm community building activities. Some biodynamic farms use the Community Supported Agriculture model, which has connections with social threefolding.

Compared to non-organic agriculture, BD farming practices have been found to be more resilient to environmental challenges, to foster a diverse biosphere, and to be more energy efficient, factors Eric Lichtfouse describes being of increasing importance in the face of climate change, energy scarcity and population growth.

Biodynamic Preparations

In his "agricultural course" Steiner prescribed nine different preparations to aid fertilization, and described how these were to be prepared. Steiner believed that these preparations mediated terrestrial and cosmic forces into the soil. The prepared substances are numbered 500 through 508, where the first two are used for preparing fields, and seven are used for making compost. A long term trial (DOK experiment) evaluating the biodynamic farming system in comparison with organic and conventional farming systems, found that both organic farming and biodynamic farming resulted in enhanced soil properties, but had lower yields than conventional farming. Regarding compost development beyond accelerating the initial phase of composting, some positive effects have been noted:

- The field sprays contain substances that stimulate plant growth include cytokinins.
- Some improvement in nutrient content of compost.

Although the preparations have direct nutrient values, their purpose in biodynamics is to support the self-regulating capacities of the soil biota in the case of 500 and 501 and the biological life resident in the composting organics, as well as the mature compost itself, in the others.

Field Preparations

Field preparations, for stimulating humus formation:

- **500**: (horn-manure) a humus mixture prepared by filling the horn of a cow with cow manure and burying it in the ground (40–60 cm below the surface) in the autumn. It is left to decompose during the winter and recovered for use the following spring.
- **501**: Crushed powdered quartz prepared by stuffing it into a horn of a cow and buried into the ground in spring and taken out in autumn. It can be mixed with 500 but usually prepared on its own (mixture of 1 tablespoon of quartz powder to 250 liters of water) The mixture is sprayed under very low pressure over the crop during the wet season, in an attempt to prevent fungal diseases. It should be sprayed on an overcast day or early in the morning to prevent burning of the leaves.

The application rate of the biodynamic field spray preparations (i.e., **500** and **501**) are 300 grams per hectare of horn manure and 5 grams per hectare of horn silica. These are made by stirring the ingredients into 20-50 liters of water per hectare for an hour, using a prescribed method.

Compost Preparations

Compost preparations, used for preparing compost, employ herbs which are frequently used in medicinal remedies. Many of the same herbs are used in organic practices to make foliar fertilizers, turned into the soil as green manure, or in composting. The preparations include:

- **502**: Yarrow blossoms (*Achillea millefolium*) are stuffed into urinary bladders from Red Deer (*Cervus elaphus*), placed in the sun during summer, buried in earth during winter and retrieved in the spring.
- **503**: Chamomile blossoms (*Matricaria recutita*) are stuffed into small intestines from cattle buried in humus-rich earth in the autumn and retrieved in the spring.
- **504**: Stinging nettle (*Urtica dioica*) plants in full bloom are stuffed together underground surrounded on all sides by peat for a year.
- **505**: Oak bark (*Quercus robur*) is chopped in small pieces, placed inside the skull of a domesticated animal, surrounded by peat and buried in earth in a place where lots of rain water runs past.
- **506**: Dandelion flowers (*Taraxacum officinale*) are stuffed into the mesentery of a cow and buried in earth during winter and retrieved in the spring.
- **507**: Valerian flowers (*Valeriana officinalis*) are extracted into water.
- **508**: Horsetail (*Equisetum*).

The compost preparations are applied with quantities of 1–2 cm^3 each per 10 m^3 compost, farmyard manure or liquid manure. The preparations should then be evenly sprayed out on the land as soon as possible after stirring.

One to three grams (a teaspoon) of each preparation is added to a dung heap by digging 50 cm deep holes with a distance of 2 meters from each other, except for the 507 preparation, which is stirred into 5 liters of water and sprayed over the entire compost surface. All preparations are thus used in homeopathic quantities. Each compost preparation is designed to guide a particular decomposition process in the composting mass. One study found that the oak bark preparation improved disease resistance in zucchini.

Planting Calendar

The approach considers that there are lunar and astrological influences on soil and plant development—for example, choosing to plant, cultivate or harvest various crops based on both the phase of the moon and the zodiacal constellation the moon is passing through, and also depending on whether the crop is the root, leaf, flower, or fruit of the plant. This aspect of biodynamics has been termed "astrological" in nature.

Seed Production

Biodynamic agriculture has focused on the open pollination of seeds (with farmers thereby generally growing their own seed) and the development of locally adapted varieties. The seed stock is not controlled by large, multinational seed companies.

Biodynamic Certification

The biodynamic certification Demeter, created in 1924, was the first certification and labelling system for organic production.[5] To receive certification as a biodynamic farm, the farm must meet the following standards: agronomic guidelines, greenhouse management, structural components, livestock guidelines, and post harvest handling and processing procedures.

The term *Biodynamic* is a trademark held by the Demeter association of biodynamic farmers for the purpose of maintaining production standards used both in farming and processing foodstuffs.(This is not a trademark held privately in New Zealand) The trademark is intended to protect both the consumer and the producers of biodynamic produce. Demeter International an organization of member countries; each country has its own Demeter organization which is required to meet international production standards (but can also exceed them). The original Demeter organization was founded in 1928; the U.S. Demeter Association was formed in the 1980s and certified its first farm in 1982. In France, Biodivin certifies biodynamic wine. In Egypt, SEKEM has created the Egyptian Biodynamic Association (EBDA), an association that provides training for farmers to become certified. As of 2006, more than 200 wineries worldwide were certified as biodynamic; numerous other wineries employ biodynamic methods to a greater or lesser extent.

Effectiveness

Research into biodynamic farming has been complicated by the difficulty of isolating the distinctively biodynamic aspects when conducting comparative trials. Consequently, there is no strong body of material that provides evidence of any specific effect.

Since biodynamic farming is a form of organic farming, it can be generally assumed to share its characteristics, including "less stressed soils and thus diverse and highly interrelated soil communities".

A 2009/2011 review found that biodynamically cultivated fields:

- had lower absolute yields than conventional farms, but achieved better efficiency of production relative to the amount of energy used;
- had greater earthworm populations and biomass than conventional farms.

Both factors were similar to the result in organically cultivated fields.

Reception

In a 2002 newspaper editorial, Peter Treue, agricultural researcher at the University of Kiel, characterized biodynamics as pseudoscience and argued that similar or equal results can be obtained using standard organic farming principles. He wrote that some biodynamic preparations more resemble alchemy or magic akin to geomancy.

In a 1994 analysis, Holger Kirchmann, a soil researcher with the Swedish University of Agricultural Sciences, concluded that Steiner's instructions were occult and dogmatic, and cannot contribute to the development of alternative or sustainable agriculture. According to Kirchmann, many of Steiner's statements are not provable because scientifically clear hypotheses cannot be made from his descriptions. Kirchmann asserted that when methods of biodynamic agriculture were tested scientifically, the results were unconvincing. Further, in a 2004 overview of biodynamic agriculture, Linda Chalker-Scott, a researcher at Washington State University, characterized biodynamics as pseudoscience, writing that Steiner did not use scientific methods to formulate his theory of biodynamics, and that the later addition of valid organic farming techniques has "muddled the discussion" of Steiner's original idea. Based on the scant scientific testing of biodynamics, Chalker-Scott concluded "no evidence exists" that homeopathic preparations improve the soil.

In Michael Shermer's *The Skeptic Encyclopedia of Pseudoscience*, Dan Dugan says that the way biodynamic preparations are supposed to be implemented are formulated solely on the basis of Steiner's "own insight". Skeptic Brian Dunning writes "the best way to think of 'biodynamic agriculture' would be as a magic spell cast over an entire farm. Biodynamics sees an entire farm as a single organism, with something that they call a life force."

Florian Leiber, Nikolai Fuchs and Hartmut Spieß, researchers at the Goetheanum, have defended the principles of biodynamics and suggested that critiques of biodynamic agriculture which deny it scientific credibility are "not in keeping with the facts...as they take no notice of large areas of biodynamic management and research." Biodynamic farmers are "charged with developing a continuous dialogue between biodynamic science and the natural sciences *sensu stricto*," despite important differences in paradigms, world views, and value systems.

Philosopher of science Michael Ruse has written that followers of biodynamic agriculture rather enjoy the scientific marginalisation that comes from its pseudoscientific basis, revelling both in its esoteric aspects and the impression that they were in the vanguard of the wider anti-science sentiment that has grown in opposition to modern methods such as genetic modification.

CLIMATE-FRIENDLY GARDENING

Climate-friendly gardening is gardening in ways which reduce emissions of greenhouse gases from gardens and encourage the absorption of carbon dioxide by soils and plants in order to aid the reduction of global warming. To be a climate-friendlygardener means considering both what happens in a garden and the materials brought into it and the impact they have on land use and climate. It can also include garden features or activities in the garden that help to reduce greenhouse gas emissions elsewhere.

Orchard garden showing orchard trees, herbaceous perennials and ground-cover plants, at Hergest Croft Gardens, Herefordshire, Britain.

Land use and Greenhouse Gases

Most of the excess greenhouse gases causing climate change have come from burning fossil fuel. But a special report from the Intergovernmental Panel on Climate Change (IPCC) estimated that in the last 150 years fossil fuels and cement production were responsible for only about two-thirds of climate change: the other third has been caused by human land use. The three main greenhouse gases produced by unsustainable land use are carbon dioxide, methane, and nitrous oxide.Black carbon or soot can also be caused by unsustainable land use, and, although not a gas, can behave like greenhouse gases and contribute to climate change.

Carbon Dioxide

Carbon dioxide, CO_2, is a natural part of the carbon cycle, but human land uses often add more, especially from habitat destruction and the cultivation of soil. When woodlands, wetlands, and other natural habitats are turned into pasture, arable fields, buildings and roads, the carbon held in the soil and vegetation becomes extra carbon dioxide and methane to trap more heat in the atmosphere.

Gardeners may cause extra carbon dioxide to be added to the atmosphere in several ways:

- Using peat or potting compost containing peat;
- Buying garden furniture or other wooden products made from woodland which has been destroyed rather than taken as a renewable crop from sustainably managed woodland;
- Digging soil and leaving it bare so that the carbon in soil organic matter is oxidised;
- Using power tools which burn fossil fuel or electricity generated by burning fossil fuel;
- Using patio heaters;
- Heating greenhouses by burning fossil fuel or electricity generated by burning fossil fuel;
- Burning garden prunings and weeds on a bonfire;
- Buying tools, pesticides, synthetic nitrogen fertilizers (over 2 kilograms of carbon dioxide equivalent is produced in the manufacture of each kilogram of ammonium nitrate), and other materials which have been manufactured using fossil fuel;
- Heating and treating swimming pools by burning fossil fuel or electricity generated by burning fossil fuel;
- Watering their gardens with tapwater, which has been treated and pumped by burning fossil fuel, with a greenhouse gas impact of about 1 kg CO_2e/m^3 water.

Gardeners will also be responsible for extra carbon dioxide when they buy garden products which have been transported by vehicles powered by fossil fuel.

Methane

Methane, CH_4, is a natural part of the carbon cycle, but human land uses often add more, especially from anaerobic soil, artificial wetlands such as rice fields, and from the guts of farm animals, especially ruminants such as cattle and sheep.

Gardeners may cause extra methane to be added to the atmosphere in several ways:

- Compacting soil so that it becomes anaerobic, for example by treading on soil when it is wet;
- Allowing compost heaps to become compacted and anaerobic;
- Creating homemade liquid feed by putting the leaves of plants such as comfrey

under water, with the unintended consequence that the plants may release methane as they decay;

- Killing pernicious weeds by covering them with water, with the unintended consequence that the plants may release methane as they decay;
- Allowing ponds to become anaerobic, for example by adding unsuitable fish species which stir up sediment that then blocks light from and kills submerged oxygenating plants.

Nitrous Oxide

Nitrous oxide, N_2O, is a natural part of the nitrogen cycle, but human land uses often add more.

Gardeners may cause extra nitrous oxide to be added to the atmosphere by:

- Using synthetic nitrogen fertilizer, for example "weed and feed" on lawns, especially if it is applied when plants are not actively growing, the soil is compacted, or when other factors are limiting so that the plants cannot make use of the nitrogen;
- Compacting the soil (for example by working in the garden when the soil is wet) which will increase the conversion of nitrates to nitrous oxide by soil bacteria;
- Burning garden waste on bonfires.

Black carbon

Black carbon is not a gas, but it acts like a greenhouse gas because it can be suspended in the atmosphere and absorb heat.

Gardeners may cause extra black carbon to be added to the atmosphere by burning garden prunings and weeds on bonfires, especially if the waste is wet and becomes black carbon in the form of soot. Gardeners will also be responsible for extra black carbon produced when they buy garden products which have been transported by vehicles powered by fossil fuel especially the diesel used in most lorries.

Gardening to Reduce Greenhouse Gas Emissions and Absorb Carbon Dioxide

There are many ways in which climate-friendly gardeners may reduce their contribution to climate change and help their gardens absorb carbon dioxide from the atmosphere.

Climate-friendly gardeners can find good ideas in many other sustainable approaches:

- Agroforestry;
- Forest gardening;
- Orchards;
- Organic gardening;
- Permaculture;

- Rain garden;
- Vegan organic gardening;
- Water-wise gardening;
- Wildlife garden.

Protecting and Enhancing Carbon Stores

Protecting Carbon Stores in Land Beyond Gardens

Woodland and wetland in the New Forest, Hampshire.

Woodland and trees in Herefordshire.

Kitchen garden at Charles Darwin's home, Down House, Kent, showing greenhouse, waterbutt, box hedging and vegetable beds.

Alliums, lavender, box and other water-thrifty plants in the dry garden at Cambridge Botanic Garden.

Climate-friendly gardening includes actions which protect carbon stores beyond gardens. The biggest carbon stores in land are in soil; the two habitat types with the biggest carbon stores per hectare are woods and wetlands; and woods absorb more carbon dioxide per hectare per year than most other habitats. Climate-friendly gardeners therefore aim to ensure that nothing they do will harm these habitats.

According to Morison and Morecroft (eds.)'s *Plant Growth and Climate Change*, the net primary productivity (the net amount of carbon absorbed each year) of various habitats is:

- Tropical forests: 12.5 tonnes of carbon per hectare per year;
- Temperate forests: 7.7 tonnes of carbon per hectare per year;
- Temperate grasslands: 3.7 tonnes of carbon per hectare per year;
- Croplands: 3.1 tonnes of carbon per hectare per year.

The Intergovernmental Panel on Climate Change's Special Report *Land Use, Land-Use Change and Forestry* lists the carbon contained in different global habitats as:

- Wetlands: 643 tonnes carbon per hectare in soil + 43 tonnes carbon per hectare in vegetation = total 686 tonnes carbon per hectare;
- Tropical forests: 123 tonnes carbon per hectare in soil + 120 tonnes carbon per hectare in vegetation = total 243 tonnes carbon per hectare;
- Temperate forests: 96 tonnes carbon per hectare in soil + 57 tonnes carbon per hectare in vegetation = total 153 tonnes carbon per hectare;
- Temperate grasslands: 164 tonnes carbon per hectare in soil + 7 tonnes carbon per hectare in vegetation = total 171 tonnes carbon per hectare;
- Croplands: 80 tonnes carbon per hectare in soil + 2 tonnes carbon per hectare in vegetation = total 82 tonnes carbon per hectare.

The figures quoted above are global averages. More recent research in 2009 has found that the habitat with the world's highest known total carbon density - 1,867 tonnes of carbon per hectare - is temperate moist forest of Eucalyptus regnans in the Central Highlands of south-east Australia; and, in general, that temperate forests contain more carbon than either boreal forests or tropical forests.

Carbon Stores in Britain

According to Milne and Brown's 1997 paper "Carbon in the vegetation and soils of Great Britain", Britain's vegetation and soil are estimated to contain 9952 million tonnes of carbon, of which almost all is in the soil, and most in Scottish peatland soil:

- Soils in Scotland: 6948 million tonnes carbon;
- Soils in England and Wales: 2890 million tonnes carbon;
- Vegetation in British woods and plantations (which cover only 11% of Britain's land area): 91 million tonnes carbon;
- Other vegetation: 23 million tonnes carbon.

A 2005 report suggested that British woodland soil may contain as much as 250 tonnes of carbon per hectare.

Many studies of soil carbon only study the carbon in the top 30 centimetres, but soil is often much deeper than that, especially below woodland. One 2009 study of the United Kingdom's carbon stores by Keith Dyson and others gives figures for soil carbon down to 100 cm below the habitats, including "Forestland", "Cropland" and "Grassland", covered by the Kyoto Protocol reporting requirements.

- Forestland soils: average figures in tonnes carbon per hectare are 160 (England), 428 (Scotland), 203 (Wales), and 366 (Northern Ireland).
- Grassland soils: average figures in tonnes carbon per hectare are 148 (England), 386 (Scotland), 171 (Wales), and 304 (Northern Ireland).
- Cropland soils: average figures in tonnes carbon per hectare are 110 (England), 159 (Scotland), 108 (Wales), and 222 (Northern Ireland).

Protecting Carbon Stores in Wetland

Permeable paving of wood chip with birch-log edging at the Royal Horticultural Society garden at Wisley.

A ground-cover and rain-garden plant - Symphytum grandiflorum, creeping comfrey (with Cotinus coggygria).

Climate-friendly gardeners choose peat-free composts because some of the planet's biggest carbon stores are in soil, and especially in the peatland soil of wetlands.

The Intergovernmental Panel on Climate Change's Special Report *Land Use, Land-Use Change and Forestry* gives a figure of 2011 gigatonnes of carbon for global carbon stocks in the top 1 metre of soils, much more than the carbon stores in the vegetation or the atmosphere.

Climate-friendly gardeners also avoid using tapwater not only because of the greenhouse gases emitted when fossil fuels are burnt to treat and pump water, but because if water is taken from wetlands then carbon stores are more likely to be oxidised to carbon dioxide.

A climate-friendly garden therefore does not contain large irrigated lawns, but instead includes water-butts to collect rainwater; water-thrifty plants which survive on rainwater and do not need watering after they are established; trees, shrubs and hedges to shelter gardens from the drying effects of sun and wind; and groundcover plants and organic mulch to protect the soil and keep it moist.[p. 242p. 80–82]

Climate-friendly gardeners will ensure that any paved surfaces in their gardens (which are kept to a minimum to increase carbon stores) are permeable, and may also make rain gardens, sunken areas into which rainwater from buildings and paving is directed, so that the rain can then be fed back into groundwater rather than going into storm drains. The plants in rain gardens must be able to grow in both dry and wet soils.

Protecting Carbon Stores in Woodland

Wetlands may store the most carbon in their soils, but woods store more carbon in their living biomass than any other type of vegetation, and their soils store the most carbon after wetlands. Climate-friendly gardeners therefore ensure that any wooden products they buy, such as garden furniture, have been made of wood from sustainably managed woodland.

Protecting and Increasing Carbon Stores in Gardens

Juglans elaeopyren, an American walnut, at Cambridge Botanic Garden.

After rocks containing carbonate compounds, soil is the biggest store of carbon on land. Carbon is found in soil organic matter, including living organisms (plant roots, fungi, animals, protists, bacteria), dead organisms, and humus. One study of the environmental benefits of gardens estimates that 86% of carbon stores in gardens is in the soil.

Wild strawberries in flower below a British hedge.\

The first priorities for climate-friendly gardeners are, therefore, to:

- Protect the soil's existing carbon stores;
- Increase the soil's carbon stores.

To protect the soil, climate-friendly gardens:

- Are based on plants rather than buildings and paving;
- Have soil that is kept at a relatively stable temperature by shelter from trees, shrubs and/or hedges;
- Have soil that is always kept covered and therefore moist and at a relatively stable temperature by groundcover plants, fast-growing green manures (which can be used as an intercrop in kitchen gardens of annual vegetables) and/or organic mulches.

Climate-friendly gardeners avoid things which may harm soil. They do not tread on the soil when it is wet, because it is then most vulnerable to compaction. They dig as little is possible, and only when the soil is moist rather than wet, because cultivation increases the oxidation of soil organic matter and produces carbon dioxide.

To increase soil carbon stores, climate-friendly gardeners ensure that their gardens create optimal conditions for vigorous healthy growth of plants, and other garden organisms above and below ground, and reduce the impact of any limiting factors.

Mulch of woodchips protecting soil at the Royal Horticultural Society garden at Wisley in Surrey.

In general, the more biomass that the plants can create each year, the more carbon will be added to the soil. However, only some biomass each year becomes long-term soil carbon or humus. In *Soil Carbon and Organic Farming*, a 2009 report for the Soil Association, Gundula Azeez discusses several factors which increase how much biomass is turned into humus. These include good soil structure, soil organisms such as fine root hairs, microorganisms, mycorrhizas and earthworms which increase soil aggregation, residues from plants (such as trees and shrubs) which have a high content of resistant chemicals such as lignin, and plant residues with a carbon to nitrogen ratio lower than about 32:1.

Nitrogen-fixing nodules on Wisteria roots (hazelnut for scale)

Climate-friendly gardens therefore include:

- Hedges for shelter from wind;
- A light canopy of late-leafing deciduous trees to let in enough sunlight for growth but not so much that the garden becomes too hot and dry (this is one of the principles behind many agroforestry systems, such as Paulownia's use in China partly because it is late-leafing and its canopy is sparse so that crops below it get shelter but also enough light);
- Groundcover plants and organic mulches (such as woodchips over compost made from kitchen and garden "waste") to keep soil moist and at relatively stable temperatures;
- Nitrogen-fixing plants, because soil nitrogen may be a limiting factor (but climate-friendly gardeners avoid synthetic nitrogen fertilizers, because these may cause mycorrhizal associations to break down);
- Many layers of plants, including woody plants such as trees and shrubs, other perennials, groundcover plants, deep-rooted plants, all chosen according to

'right plant, right place', so that they are suited to their growing conditions and will grow well;

- A wide diversity of disease-resistant, vigorous plants for resilience and to make the most of all available ecological niches;
- Plants to feed and shelter wildlife, to increase total biomass, and to ensure biological control of pests and diseases.
- Compost made from garden and kitchen "waste".

Lawns, like other grasslands, can build up good levels of soil carbon, but they will grow more vigorously and store more carbon if besides grasses they also contain nitrogen-fixing plants such as clover, and if they are cut using a mulching mower which returns finely-chopped mowings to the lawn. More carbon, however, may be stored by other perennial plants such as trees and shrubs. They also do not need to be maintained using power tools.

Climate-friendly gardeners will also aim to increase biodiversity not only for the sake of the wildlife itself, but so that the garden ecosystem is resilient and more likely to store as much carbon as possible as long as possible. They will therefore avoid pesticides, and increase the diversity of the habitats within their gardens.

Reducing Greenhouse Gas Emissions

Climate-friendly gardeners can directly reduce the greenhouse gas emissions from their own gardens, but can also use their gardens to indirectly reduce greenhouse gas emissions elsewhere.

Using gardens to Reduce Greenhouse Gas Emissions

Climate-friendly gardeners can use their gardens in ways which reduce greenhouse gases elsewhere, for example by using the sun and wind to dry washing on washing lines in the garden instead of using electricity generated by fossil fuel to dry washing in tumble dryers.

From Farmland

Food is a major contributor to climate change. In the United Kingdom, according to Tara Garnett of the Food Food Climate Research Network, food contributes 19% of the country's greenhouse gas emissions.

Soil is the biggest store of carbon on land. It is therefore important to protect the soil organic matter in farmland. Farm animals, however, especially free-range pigs, may cause erosion, and cultivation of the soil increases the oxidation of soil organic matter into carbon dioxide. Other sources of greenhouse gases from farmland include: compaction caused by farm machinery or overgrazing by farm animals can make soil anaerobic and produce methane; farm animals produce methane; and nitrogen fertilizers can be converted to nitrous oxide. Most farmland consists of fields growing annual arable crops which are eaten directly by people or fed to farm animals, and grassland used as pasture, hay or silage to feed farm animals. Some perennial food plants are also grown, such as fruits and nuts in orchards, and watercress grown in water.

Walnut, Juglans regia, with ripening walnuts.

Although all cultivation of the soil in arable fields produces carbon dioxide, some arable crops cause more damage to soil than others. Root crops such as potatoes and sugar-beet, and crops which are harvested not just once a year but over a long period such as green vegetables and salads, are considered "high risk" in catchment-sensitive farming.

Climate-friendly gardeners therefore grow at least some of their food, and may choose food crops which therefore help to keep carbon in farmland soils if they grow such high-risk crops in small vegetable plots in their gardens, where it is easier to protect the soil than in large fields under commercial pressures. Climate-friendly gardeners may grow and eat plants such as sweet cicely which sweeten food, and so reduce the land area needed for sugar-beet. They may also choose to grow perennial food plants to not only reduce their indirect greenhouse gas emissions from farmland, but also to increase carbon stores in their own gardens.

Grassland contains more carbon per hectare than arable fields, but farm animals, especially ruminants such as cattle or sheep, produce large amounts of methane, directly and from manure heaps and slurry. Slurry and manure may also produce nitrous oxide. Gardeners who want to reduce their greenhouse gas emissions can help themselves to eat less meat and dairy produce by growing nut trees which are a good source of tasty, protein-rich food, including walnuts which are an excellent source of the omega-3 fatty acid alpha-linolenic acid.

Researchers and farmers are investigating and improving ways of farming which are more sustainable, such as agroforestry, forest farming, wildlife-friendly farming, soil management, catchment-sensitive farming (or water-friendly farming). For example, the organisation Farming Futures assists farmers in the United Kingdom to reduce their farms' greenhouse gas emissions.

Farmers are aware that consumers are increasingly asking for green credentials. Gardeners who understand climate-friendly practices can advocate their use by farmers.

From Industry

Nitrogen-fixing and edible - Elaeagnus umbellatus at the Agroforestry Research Trust forest garden in Devon.

Climate-friendly gardeners aim to reduce their consumption in general. In particular, they try to avoid or reduce their consumption of tapwater because of the greenhouse gases emitted when fossil fuels are burnt to supply the energy needed to treat and pump it to them. Instead, gardeners can garden using only rainwater.

Greenhouse gases are produced in the manufacture of many materials and products used by gardeners. For example, it takes a lot of energy to produce synthetic fertilizers, especially nitrogen fertilizers. Ammonium nitrate, for example, has an embodied energy of 67000 kilojoules/kilogramme, so climate-friendly gardeners will choose alternative ways of ensuring the soil in their gardens has optimal levels of nitrogen by alternative means such as nitrogen-fixing plants.

Climate-friendly gardeners will also aim to follow "cradle-to-cradle design" and "circular economy" principles: when they choose to buy or make something, it should be possible to take it apart again and recycle or compost every part, so that there is no waste, only raw materials to be made into something else. This will reduce the greenhouse gases otherwise produced when extracting raw materials.

From Transport

Gardeners can reduce not only their food miles by growing some of their own food, but also their "gardening miles" by reducing the amount of plants and other materials they import, obtaining them as locally as possible and with as little packaging as possible. This might include ordering plants by mail order from a specialist nursery if the plants are sent out bare-root, reducing transport demand and the use of peat-based composts; or growing plants from seed, which will also increase genetic diversity and therefore resilience; or growing plants vegetatively from cuttings or offsets from other local gardeners; or buying reclaimed materials from salvage firms.

From Houses

Climbers as insulation - Boston ivy, Parthenocissus tricuspidata, Boston ivy, in autumn.

Climate-friendly gardeners can use their gardens in ways which reduce greenhouse gas emissions from homes by:

- Using sunlight and wind to dry washing on washing lines instead of fossil fuel-generated electricity to run tumble dryers;
- Planting deciduous climbers on houses and planting deciduous trees at suitable distances from the house to provide shade during the summer, reducing the consumption of electricity for air conditioning, but also such that at cooler times of year, sunlight can reach and warm a house, reducing heating costs and consumption;
- Planting hedges, trees, shrubs and climbers to shelter houses from wind, reducing heating costs and consumption during the winter (as long as any planting does not create a wind-tunnel effect).

Climate-friendly gardeners may also choose to reduce their own personal greenhouse gas emissions by growing and eating carminative plants such as fennel and garlic which reduce intestinal gases such as methane.

Reducing Greenhouse Gas Emissions from Gardens

Slow-growing yew, Taxus baccata, as hedge at Charles Darwin's home, Down House, Kent.

Nitrogen-fixing red and white clover (Trifolium) as lawn plants.

Leaf cage, compost heap and wormery at the Royal Horticultural Society garden at Wisley.

There are some patent sources of greenhouse gas emissions in gardens and some more latent.

Power tools which are powered by diesel or petrol, or electricity generated by burning other fossil fuels, emit carbon dioxide. Climate-friendly gardeners may therefore choose to use hand tools rather than power tools, or power tools powered by renewable electricity, or design their gardens to reduce or remove a need to use power tools. For example, they may choose dense, slow-growing species for hedges so that the hedges only need to be cut once a year.

Lawns need[?] to be cut by lawn mowers and, in drier parts of the world, are often irrigated by tapwater. Climate-friendly gardeners will therefore do what they can to reduce this consumption by:

- Replacing part of or all lawns with other perennial planting such as trees and shrubs with less ecologically demanding maintenance requirements;
- Cut some or all lawns only once or twice a year, i.e. convert them into meadows;
- Make lawn shapes simple so that they may be cut quickly;
- Increase the cutting height of mower blades;
- Use a mulching mower to return organic matter to the soil;
- Sow clover to increase vigour (without the need for synthetic fertilisers) and resilience in dry periods;
- Cut lawns with electric mowers using electricity from renewable energy;
- Cut lawns with hand tools such as push mowers or scythes.

Greenhouses can be used to grow crops which might otherwise be imported from warmer climates, but if they are heated by fossil fuel then they may cause more greenhouse gas emissions than they save. Climate-friendly gardeners will therefore use their greenhouses carefully by:

- Choosing only annual plants which will only be in the greenhouse during warmer months, or perennial plants which do not need any extra heat during winter;
- Using water tanks as heat stores and compost heaps as heat sources inside greenhouses so that they stay frost-free in winter.

Climate-friendly gardeners will not put woody prunings on bonfires, which will emit carbon dioxide and black carbon, but instead burn them indoors in a wood-burning stove and therefore cut emissions from fossil fuel, or cut them up to use as mulch and increase soil carbon stores, or add the smaller prunings to compost heaps to keep them aerated, reducing methane emissions. To reduce the risk of fire, they will also choose fire-resistant plants from habitats which are not prone to wildfires and which do not catch fire easily, rather than fire-adapted plants from fire-prone habitats which are flammable and adapted to encourage fires and then gain a competitive advantage over less resistant species.

Climate-friendly gardeners may use deep-rooted plants such as comfrey to bring nutrients closer to the surface topsoil, but will do so without making the leaves into a liquid feed, because the rotting leaves in the anaerobic conditions under water may emit methane.

Nitrogen fertilizers may be oxidised to nitrous oxide, especially if fertilizer is applied in excess, or when plants are not actively growing. Climate-friendly gardeners may choose instead to use nitrogen-fixing plants which will add nitrogen to the soil without increasing nitrous oxide emissions.

Forest Gardening

Robert Hart's forest garden in Shropshire.

Forest gardening is a low-maintenance sustainable plant-based food production and agroforestry system based on woodland ecosystems, incorporating fruit and nut trees, shrubs, herbs, vines and perennial vegetables which have yields directly useful to humans. Making use of companion planting, these can be intermixed to grow in a succession of layers, to build a woodland habitat.

Forest gardening is a prehistoric method of securing food in tropical areas. In the 1980s, Robert Hart coined the term forest gardening after adapting the principles and applying them to temperate climates.

Forest gardens are probably the world's oldest form of land use and most resilient agroecosystem. They originated in prehistoric times along jungle-clad river banks and in the wet foothills of monsoon regions. In the gradual process of families improving their immediate environment, useful tree and vine species were identified, protected and improved whilst undesirable species were eliminated. Eventually superior foreign species were selected and incorporated into the gardens.

Forest gardens are still common in the tropics and known by various names such as: *home gardens* in Kerala in South India, Nepal, Zambia, Zimbabwe and Tanzania; *Kandyan forest gardens* in Sri Lanka; *huertos familiares*, the "family orchards" of Mexico; and *pekarangan*, the gardens of "complete design", in Java. These are also called agroforests and, where the wood components are short-statured, the term shrub garden is employed. Forest gardens have been shown to be a significant source of income and food security for local populations.

Robert Hart adapted forest gardening for the United Kingdom's temperate climate during the 1980s. His theories were later developed by Martin Crawford from the Agroforestry Research Trust and various permaculturalists such as Graham Bell, Patrick Whitefield, Dave Jacke and Geoff Lawton.

In Tropical Climates

Forest gardens, or home gardens, are common in the tropics, using intercropping to cultivate trees, crops, and livestock on the same land. In Kerala in south India as well as in northeastern India, the home garden is the most common form of land use and is also found in Indonesia. One example combines coconut, black pepper, cocoa and pineapple. These gardens exemplify polyculture, and conserve much crop genetic diversity and heirloom plants that are not found in monocultures. Forest gardens have been loosely compared to the religious concept of the Garden of Eden.

Americas

The BBC's Unnatural Histories claimed that the Amazon rainforest, rather than being a pristine wilderness, has been shaped by humans for at least 11,000 years through practices such as forest gardening and *terra preta*. This was also explored in the bestselling book *1491* by author Charles C. Mann. Since the 1970s, numerous geoglyphs have also been discovered on deforested land in the Amazon rainforest, furthering the evidence about Pre-Columbian civilizations.

On the Yucatán Peninsula, much of the Maya food supply was grown in orchard-gardens, known as *pet kot*. The system takes its name from the low wall of stones (*pet* meaning circular and *kot* wall of loose stones) that characteristically surrounds the gardens.

Africa

In many African countries, for example Zambia, Zimbabwe, Ethiopia and Tanzania, gardens are widespread in rural, periurban and urban areas and they play an essential role in establishing food security. Most well known are the Chaga or Chagga gardens on the slopes of Mt. Kilimanjaro in Tanzania. These are an excellent example of an agroforestry system. In many countries, women are the main actors in home gardening and food is mainly produced for subsistence. In North-Africa, oasis layered gardening with palm trees, fruit trees and vegetables is a traditional type of forest garden.

Nepal

In Nepal, the *Ghar Bagaincha*, literally "home garden", refers to the traditional land use system around a homestead, where several species of plants are grown and maintained by household members and their products are primarily intended for the family consumption (Shrestha et al., 2002). The term "home garden" is often considered synonymous to the kitchen garden. However, they differ in terms of function, size, diversity, composition and features (Sthapit et al., 2006). In Nepal, 72% of households have home gardens of an area 2–11% of the total land holdings (Gautam et al., 2004). Because of their small size, the government has never identified home gardens as an important unit of food production and they thereby remain neglected from research and development. However, at the household level the system is very important as it is an important source of quality food and nutrition for the rural poor and, therefore, are important contributors to the household food security and livelihoods of farming communities in Nepal. The gardens are typically cultivated with a mixture of annual and perennial plants that can be harvested on a daily or seasonal basis. Biodiversity that has an immediate value is maintained in home gardens as women and children have easy access to preferred food. Home gardens, with their intensive and multiple uses, provide a safety net for households when food is scarce. These gardens are not only important sources of food, fodder, fuel, medicines, spices, herbs, flowers, construction materials and income in many countries, they are also important for the in situ conservation of a wide range of unique genetic resources for food and agriculture (Subedi et al., 2004). Many uncultivated, as well as neglected and underutilised species could make an important contribution to the dietary diversity of local communities (Gautam et al., 2004).

In addition to supplementing diet in times of difficulty, home gardens promote whole-family and whole-community involvement in the process of providing food. Children, the elderly, and those caring for them can participate in this infield agriculture, incorporating it with other household tasks and scheduling. This tradition has existed in many cultures around the world for thousands of years.

In Mediterranean climates

The Mediterranean climate has long, hot, rainless summers and relatively short, cool, rainy winters (Köppen climate classification *Csa*). Its climate conditions are highly variable within an area and modified locally by altitude, latitude, and the proximity to the Mediterranean. In the 1950s the Forest Research Department of the Ministry of Agriculture founded a botanical forest garden in the Sharon region in Israel, the Ilanot Forest. As the only one of its kind in Israel, it harbours more than 750 species of trees from locations all over the world, including the Japanese sago palm cycas revoluta, fig trees (ficus glomerata), stone pine trees (pinus pinea) that produce tasty pine nuts and adds to the biodiversity of Israel.

In Temperate Climates

Robert Hart, forest gardening pioneer.

Robert Hart coined the term "forest gardening" during the 1980s. Hart began farming at Wenlock Edge in Shropshire with the intention of providing a healthy and therapeutic environment for himself and his brother Lacon. Starting as relatively conventional smallholders, Hart soon discovered that maintaining large annual vegetable beds, rearing livestock and taking care of an orchard were tasks beyond their strength. However, a small bed of perennial vegetables and herbs he planted was looking after itself with little intervention.

Following Hart's adoption of a raw vegan diet for health and personal reasons, he replaced his farm animals with plants. The three main products from a forest garden are fruit, nuts and green leafy vegetables. He created a model forest garden from a 0.12 acre (500 m^2) orchard on his farm and intended naming his gardening method *ecological horticulture* or *ecocultivation*. Hart later dropped these terms once he became aware that *agroforestry* and *forest gardens* were already being used to describe similar systems in other parts of the world. He was inspired by the forest farming methods of Toyohiko Kagawa and James Sholto Douglas, and the productivity of the Keralan home gardens as Hart explains:

From the agroforestry point of view, perhaps the world's most advanced country is the Indian state of Kerala, which boasts no fewer than three and a half million forest gardens...As an example of the extraordinary intensivity of cultivation of some forest gardens, one plot of only 0.12 hectares (0.30 acres) was found by a study group to have twenty-three young coconut palms, twelve cloves, fifty-six bananas, and forty-nine pineapples, with thirty pepper vines trained up its trees. In addition, the small holder grew fodder for his house-cow.

Seven-Layer System

The seven layers of the forest garden.

Robert Hart pioneered a system based on the observation that the natural forest can be divided into distinct levels. He used intercropping to develop an existing small orchard of apples and pears into an edible polyculture landscape consisting of the following layers:

- 'Canopy layer' consisting of the original mature fruit trees.
- 'Low-tree layer' of smaller nut and fruit trees on dwarfing root stocks.
- 'Shrub layer' of fruit bushes such as currants and berries.
- 'Herbaceous layer' of perennial vegetables and herbs.
- 'Rhizosphere' or 'underground' dimension of plants grown for their roots and tubers.
- 'Ground cover layer' of edible plants that spread horizontally.
- 'Vertical layer' of vines and climbers.

A key component of the seven-layer system was the plants he selected. Most of the traditional vegetable crops grown today, such as carrots, are sun loving plants not well selected for the more shady forest garden system. Hart favoured shade tolerant perennial vegetables.

Further Development

The Agroforestry Research Trust (ART), managed by Martin Crawford, runs experimental forest gardening projects on a number of plots in Devon, United Kingdom. Crawford describes a forest garden as a low-maintenance way of sustainably producing food and other household products.

Ken Fern had the idea that for a successful temperate forest garden a wider range of edible shade tolerant plants would need to be used. To this end, Fern created the organisation Plants for a Future (PFAF) which compiled a plant database suitable for such a system. Fern used the term *woodland gardening*, rather than forest gardening, in his book *Plants for a Future*.

The Movement for Compassionate Living (MCL) promote forest gardening and other types of vegan organic gardening to meet society's needs for food and natural resources. Kathleen Jannaway, the founder of MCL, wrote a book outlining a sustainable vegan future called *Abundant Living in the Coming Age of the Tree* in 1991. In 2009, the MCL provided a grant of £1,000 to the Bangor Forest Garden project in Gwynedd, North West Wales.

Kevin Bradley coined the phrase "Edible Forest" in the 1980s as the name of his nursery, garden, and orchard on 5 acres in the frigid zone 3 pine forests of northern Wisconsin. Among 3 options, he chose "Edible Forest" because it evokes at once an ethereal, spiritual, and magical image, of Disney- like Forest of No Return; of the biblical Garden of Eden. This image was perfectly in line with his ongoing experiment begun in 1985 in what he calls a closed loop human environment, combining multi- story tree and field crop garden/orchards for maximum beauty and use of space, someday to be very useful in an ever shrinking world. The name, at the same time, with its irrational first impression (of course we can't eat a forest), forces the mind to think, if just a little bit, about its inference and thus sticks in our memories. It appeared from Bradley's research that the two words had, prior to the 80's, never been put together before as a noun phrase but which by today, after more than two decades of Bradley's Edible Forest Nursery and the 2005 text by Jacke and Toensmeirer's- Edible Forest Gardens, has grown into a movement and little Edible Forests all over the world.

In 2005, Dave Jacke and Eric Toensmeier's two-volume *Edible Forest Gardens* provided a deeply researched reference focused on North American forest gardening climates, habitats, and species. The book attempts to ground forest gardening deeply in ecological science. The Apios Institute wiki grew out of their work, and seeks to document and share the experience of people around the world working with the species in polycultures.

Permaculture

Bill Mollison, who coined the term *permaculture*, visited Robert Hart at his forest garden in Wenlock Edge in October 1990.Hart's seven-layer system has since been adopted as a common permaculture design element.

Numerous permaculturalists are proponents of forest gardens, or food forests, such as Graham Bell, Patrick Whitefield, Dave Jacke, Eric Toensmeier and Geoff Lawton. Bell started building his forest garden in 1991 and wrote the book *The Permaculture Garden* in 1995, Whitefield wrote the book *How to Make a Forest Garden* in 2002, Jacke and Toensmeier co-authored the two volume book set *Edible Forest Gardening* in 2005, and Lawton presented the film *Establishing a Food Forest* in 2008.

Austrian Sepp Holzer practices "Holzer Permaculture on his *Krameterhof* farm, at varying altitudes ranging from 1,100 to 1,500 metres above sea level. His designs create micro-climates with rocks, ponds and living wind barriers, enabling the cultivation of a variety of fruit trees, vegetables and flowers in a region that averages 4 °C, and with temperatures as low as -20 °C in the winter.

Projects

El Pilar on the Belize-Guatemala border features a forest garden to demonstrate traditional Maya agricultural practices.A further 1-acre model forest garden, called Känan K'aax (meaning well-tended garden in Mayan), is being funded by the National Geographic Society and developed at Santa Familia Primary School in Cayo.

In the United States the largest known food forest on public land is believed to be the 7-acre Beacon Food Forest in Seattle, Washington. Other forest garden projects include those at the Central Rocky Mountain Permaculture Institute in Basalt, Colorado and Montview Neighborhood farm in Northampton, Massachusetts.

In Canada food forester Richard Walker has been developing and maintaining food forests in the province of British Columbia for over 30 years. He developed a 3-acre food forest that when at maturity provided raw materials for a nursery and herbalism business as well as food for his family. The Living Centre have developed various forest garden projects in Ontario.

In the United Kingdom, other than those run by the Agroforestry Research Trust (ART), there are numerous forest garden projects such as the Bangor Forest Garden in Gwynedd, North West Wales. Martin Crawford from ART administers the Forest Garden Network, an informal network of people and organisations around the world who are cultivating their own forest gardens.

SQUARE FOOT GARDENING

Square Foot Garden in Raised Bed.

Square foot gardening is the practice of dividing the growing area into small square sections (typically 12" on a side, hence the name). The aim is to assist the planning and creating of a small but intensively planted vegetable garden. It results in a simple and orderly gardening system, from which it draws much of its appeal. Mel Bartholomew coined the term "square foot gardening" in his 1981 book of the same name.

Overview

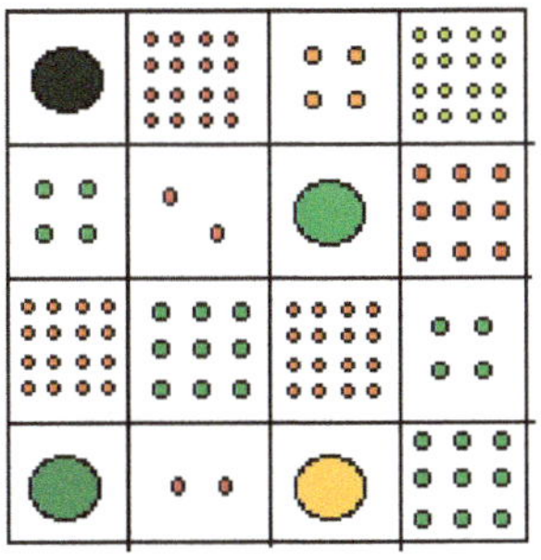

A basic, 4x4, 16-unit "square-foot garden."

The phrase "square foot gardening" was popularized by Mel Bartholomew in a 1981 Rodale Press book and subsequent PBS television series. Bartholomew used a 12' by 12' square with a grid that divided it into 9 squares with equal lengths of 4 feet on each side. Each of these 4' by 4' squares was then invisibly divided into sixteen one foot squares that were each planted with a different species. In smaller square gardens the grids may simply serve as a way to divide the garden but in larger gardens the grids can be made wide enough to be used as narrow walkways. Bartholomew recommends careful spacing of seeds rather than planting the entire seed packet so that fewer but stronger plants will grow.

To encourage a variety of different crops over time, each square would be used for a different kind of plant, the number of plants per square depending on an individual plant's size. For example, a single tomato plant might take a full square, as might herbs such as oregano, basil or mint, while most strawberry plants could be planted four per square, and up to sixteen per square of plants such as radish. Tall or climbing plants such as maize or pole beans might be planted in a northern row (south in the southern hemisphere) so as not to shade other plants, and supported with lattice or netting.

One advantage of densely planted crops is that they can form a living mulch, and also prevent weeds from establishing or even germinating. Also, natural insect repellent methods such as companion planting (i.e. planting marigolds or other naturally pest-repelling plants) become more efficient in a close space, which may reduce the need to use pesticides. The large variety of crops in a small space also prevents plant diseases from spreading easily

Since the beds are typically small, making covers or cages to protect plants from pests, cold, or sun is more practical than with larger gardens. To extend the growing season of a square foot garden, a cold/hot frame may be built around the SFG, and by facing the cold/hot frame south, the SFG captures more light and heat during the colder months of spring and winter.

In 2006, Bartholemew updated the concept with the book "All New Square Foot Gardening", which advocates growing in raised beds instead of the ground. The rationale is that by using potting soil instead of dirt, one will enjoy the benefits of having perfect soil conditions from day one, instead of needing to undertake a period of soil improvement.

VEGAN ORGANIC GARDENING

Vegan organic gardening and farming is the organic cultivation and production of food crops and other crops with a minimal amount of exploitation or harm to any animal. Vegan gardening and stock-free farming methods use no animal products or by-products, such as bloodmeal, fish products, bone meal, feces, or other animal-origin matter, because the production of these materials is viewed as either harming animals directly, or being associated with the exploitation and consequent suffering of animals. Some of these materials are by-products of animal husbandry, created during the process of cultivating animals for the production of meat, milk, skins, furs, entertainment, labor, or companionship; the sale of by-products decreases expenses and increases profit for those engaged in animal husbandry, and therefore helps support the animal husbandry industry, an outcome most vegans find unacceptable.

Types

Veganiculture

Vegan - Organic - Agriculture / Permaculture The Future Of Farming! All Things Related To: Organic Gardening, Farming & Food Forests Free From Animals & Animal Products

Forest Gardening

Forest gardening is a fully plant-based organic food production system based on woodland ecosystems, incorporating fruit and nut trees, shrubs, herbs, vines and perennial vegetables. Making use of companion planting, these can be intermixed to grow in a succession of layers, to replicate a woodland habitat. Forest gardening can be viewed as a way to recreate the Garden of Eden. The three main products from a forest garden are fruit, nuts and green leafy vegetables.

Robert Hart adapted forest gardening for temperate zones during the early 1960s. Robert Hart began with a conventional smallholding at Wenlock Edge in Shropshire. However, following his adoption of a raw vegan diet for health and personal reasons, Hart replaced his farm animals with plants. He created a model forest garden from a small orchard on his farm and intended naming his gardening method *ecological horticulture* or *ecocultivation*. Hart later dropped these terms once he became aware that *agroforestry* and *forest gardens* were already being used to describe similar systems in other parts of the world.

Vegan Permaculture

Vegan permaculture (also known as veganic permaculture, veganiculture, or vegaculture) avoids the use of domesticated animals. It is essentially the same as permaculture except for the addition of a fourth core value; "Animal Care." Zalan Glen, a raw vegan, proposes that *vegaculture* should emerge from permaculture in the same way veganism split from vegetarianism in the 1940s. Vegan permaculture recognizes the importance of free-living animals, not domesticated animals, to create a balanced ecosystem.

Veganic Gardening

The veganic gardening method is a distinct system developed by Rosa Dalziell O'Brien, Kenneth Dalziel O'Brien and May E. Bruce, although the term was originally coined by Geoffrey Rudd as a contraction of *vegetable organic* in order to "denote a clear distinction between conventional chemical based systems and organic ones based on animal manures". The O'Brien system's principal argument is that animal manures are harmful to soil health rather than that their use involves exploitation of and cruelty to animals.

The system employs very specific techniques including the addition of straw and other vegetable wastes to the soil in order to maintain soil fertility. Gardeners following the system use soil-covering mulches, and employ non-compacting surface cultivation techniques using any short-handled, wide-bladed, hand hoe. They kneel when surface cultivating, placing a board under their knees to spread out the pressure, and prevent soil compaction. Kenneth Dalziel O'Brien published a description of his system in *Veganic Gardening, the Alternative System for Healthier Crops*:

The veganic method of clearing heavily infested land is to take advantage of a plant's tendencies to move its roots nearer to the soil's surface when it is deprived of light. To make use of this principle, aided by a decaying process of the top growth of weeds, etc., it is necessary to subject such growth to heat and moisture in order to speed up the decay, and this is done by applying lime, then a heavy straw cover, and then the herbal compost activator...The following are required: Sufficient new straw to cover an area to be cleared to a depth of 3 to 4 inches.

The O'Brien method also advocates minimal disturbance of the soil by tilling, the use of cover crops and green manures, the creation of permanent raised beds and permanent hard-packed paths between them, the alignment of beds along a north-south axis, and planting in double rows or more so that not every row has a path on both sides. Use of animal manure is prohibited.

Vegan Biodynamic Agriculture

The German agricultural researcher Maria Thun (1922 - 2012) developed vegan equivalents to the traditional, animal based biodynamic preparations. As a reaction to the BSE scandal in Europe she started researching plant based preparations, using tree barks as replacement for animal organs as sheath for the preparations.

In particular in Italy, there is a movement of vegan biodynamic farming, represented by farmers such as Sebastiano Cossia Castiglioni and Cristina Menicocci.

There are many other methods currently used and under development.

Practices

Soil fertility is maintained by the use of green manures, cover crops, green wastes, composted vegetable matter, and minerals. Some vegan gardeners may supplement this with human urine from vegans (which provides nitrogen) and 'humanure' from vegans, produced from compost toilets. Generally only waste from vegans is used because of the expert recommendation that the risks associated with using composted waste are acceptable only if the waste is from animals or humans having a largely herbivorous diet.

Veganic gardeners may prepare soil for cultivation using the same method used by conventional and organic gardeners of breaking up the soil with hand tools and power tools and allowing the weeds to decompose.

Wildlife Garden

Joe-Pye weed in flower.

A **wildlife garden** (or **wild garden**) is an environment created by a gardener that serves as a sustainable haven for surrounding wildlife. Wildlife gardens contain a variety of habitats that cater to native and local plants, birds, amphibians, reptiles, insects, mammals and so on. Establishing a garden environment that mimics surrounding wildlife allows for natural systems to interact and establish an equilibrium, ultimately minimizing the need for gardener maintenance and intervention. Wildlife gardens can also play an essential role in biological pest control, and also promote biodiversity, native plantings, and generally benefit the wider environment.

Habitats

Building a successful garden suitable for local wildlife is best accomplished through the use of multiple three-dimensional habitats with diverse structures that provide places for animals to nest and hide. Wildlife gardens may contain a range of habitats, including:

Log piles – Preferably located in a shady area, a pile of logs is a sanctuary for insects and amphibians. The organic structure is a shelter for both protection and breeding. In addition to logs, garden debris may also be added around the garden to be used as a natural mulch, fertilizer, weed control, soil amendment, and habitat for arthropod predators.

Bird feeding stations and bird houses – A place for birds to eat and take shelter will increase the amount of birds in the garden, which play a key role in biological pest control. Not only will food and shelter increase the survival rate of birds, but it will also ensure that they are healthy enough for a successful breeding season.

Bug boxes – Offcuts of wood placed in a structure above ground provides an alternate place of shelter for beneficial insects, such as the robber fly, which help keep natural ecosystem predators in check.

Sources of water – A water feature, such as a pond, has the potential to support a large biodiversity of wildlife. To maximize the amount of wildlife attracted to the water feature, it should consist of ranging depths. Shallow areas are used by birds to drink and by insects and amphibians to lay eggs. Deeper areas provide habitat for aquatic insects and a place for amphibians, or even fish to swim.

Pollinating flowers – Flowers rich in nectar will attract bees and butterflies into the garden. Wildflower meadows are an alternative option for lawns in the garden and will serve as a sanctuary for pollinators. However, pollinating plants should not be confused with plants suitable for butterfly breeding.

Plant diversity – The garden should include a range of plant types to serve different specie habitats. A balance between ground cover, shrub, understory, and canopy species will allow different sized wildlife shelters that fit their individual needs.

Choice of Plants

Although some exotics may also be included, wild gardens usually mostly feature a variety of native species. Generally, these will be a part of the pre-existing natural ecology of an area, making them easier to grow than most exotic species. Choosing native plants comes with an array of benefits for both plant and animal diversity, especially the ability to support native insect and mushroom populations that have established balanced evolutionary relationships over thousands of years.

Ornamental plants on the market tend to lean toward "pest-free" plants, making it hard for native insects to adapt, and ultimately reducing their food supply. Decreases in insect populations due to excessive ornamental planting will discourage bird populations from inhabiting the area.

Invasive species can always prove problematic in the garden due to the absence of natural predators and their ability to reproduce rapidly. Without any measures of control, invasive species can easily overtake native species in the garden. Addressing invasive plants can be done a variety of ways; however, to ensure the least amount of damage to the surrounding ecosystem, this is best done by cutting down the plant, followed by painting its stem with an herbicide, such as Roundup.The debris from the invasive species can be piled and used as a home for smaller critters.

Wildlife gardens in the Netherlands

Wildlife gardens in the Netherlands are called "heemtuinen". The first was created in 1925: Thijsse's Hof (Garden of Thijsse) in Bloemendaal, near Haarlem. It was gifted to Jac. P. Thijsse on the occasion of his 60th anniversary, and still exists today. The garden gives a display of about 800 plants native to the dune region of South Kennemerland, in which the garden is situated. It is said to be one of the oldest wildlife gardens of its sort in the world.

Nowadays some 25 wildlife gardens exist in the Netherlands.

REFERENCES

1. Hix, John. 1974. The glass house. Cambridge, Mass: MIT Press.
2. Pati, Ranjan; Abelar, Michael (27 May 2015). "The Application and Optimization of Metal Reflectors to Vertical Greenhouses to Increase Plant Growth and Health". Journal of Agricultural Engineering and Biotechnology: 63–71. doi:10.18005/JAEB0302003.
3. "Glossary for Vertical Farming". Retrieved 2016-01-06.
4. Vertical farming (1915). Wilmington, Del.: E. I. duPont de Nemours Powder Co. Retrieved 2011-07-23.
5. Venkataraman, Bina (2008-07-15). "Country, the City Version: Farms in the Sky Gain New Interest". The New York Times. Retrieved 2011-01-05.
6. https://www.youtube.com/watch?v=1t-PUIXUPgw
7. https://www.youtube.com/watch?v=zWommg6JVfY
8. https://www.youtube.com/watch?v=r-Iymp-gdbo
9. https://www.youtube.com/watch?v=PrLTuGKcvPA
10. "The Rise of Urban Farming". Smithsonian Magazine. Retrieved 2011-01-05.
11. Douglas, James Sholto (1975). Hydroponics: The Bengal System(5th ed.). New Dehli: Oxford University Press. ISBN 9780195605662.
12. Nelson, Bryn (2007-12-12). "Could vertical farming be the future? Nelson, B.

(2008).". MSNBC. Retrieved 2010-11-10.

13. George Monbiot (2010-08-16). "Monbiot, G. (16 August 2010). Greens living in ivory towers now want to farm them too. 'The". London: Guardian. Retrieved 2010-11-10.
14. https://store.ctbuh.org/PDF_Previews/Journal/CTBUHJournal_2009-2.pdf
15. Ruby, Andreas. Groundscapes pp. 87–93
16. Eaton, Ruth. Ideal Cities p. 239
17. link Archived August 11, 2011, at the Wayback Machine.
18. link
19. http://www.impactlab.net/wp-content/uploads/2008/10/editt-tower-by-tr-hamzah-and-yeang.jpg
20. http://ps-survival.com/PS/Hydroponics/Hydroponics_The_Bengal_System_1959.pdf
21. Arnie Cooper (May 19, 2009). "Going up? Farming in High Rises Raises Hopes". Miller-mccune.com. Retrieved 2010-11-10.
22. Whyte, Murray (2008-07-27). "Is high rise farming in Toronto's future?". Toronto Star. Retrieved 2008-08-12.
23. "Sky Farm Proposed for Downtown Toronto". TreeHugger. Retrieved 2009-03-14.
24. Venkataraman, Bina (2008-07-15). "Country, the City Version: Farms in the Sky Gain New Interest". The New York Times.
25. Shute, Nancy (2007-05-20). "Farm of the Future? Someday food may grow in skyscrapers". U.S. News & World Report.
26. Feldman, Amy (2007-07-11). "Skyscraper Farms". Popular Science.
27. Despommier, Dickson (November 2009). "The Rise of Vertical Farms". Scientific American. New york: Scientific American Inc. 301 (5): 60–67. ISSN 0036-8733.
28. http://inhabitat.com/vertical-harvest-of-jackson-hole-will-grow-organic-produce-even-in-the-winter/
29. "Vertical Fresh Farms LLC, Buffalo, NY | StateLog". www.statelog.com. Retrieved 2015-11-05.
30. Clive Clifton says: (2009-08-24). "Vertical Farming: Too Far Outside the Box? |". E4capital.com. Retrieved 2010-11-10.
31. Banerjee C., Adenaeuer L. (2014). Up, up and away! The economics of vertical farming http://www.macrothink.org/journal/index.php/jas/article/view/4526
32. Michael Bomford says: (2010-12-02). "Energy and VerticalFarms". Retrieved 2011-03-08.
33. Evans, P. (July 22, 2009). Local food no green panacea: professor. CBC News http://www.cbc.ca/consumer/story/2009/07/22/consumer-local-food.html
34. https://www.youtube.com/watch?v=KZN3ucwePGk
35. "Does it really stack up?". The Economist. 2010-12-09.

36. "Starting a Commercial Greenhouse Business". Omafra.gov.on.ca. 2003-07-28. Retrieved 2010-11-10.

37. Pocket World in Figures, The Economist, 2011 ed. pg 64

38. Roach, J. (June 30, 2009). "High-Rise Farms: The Future of Food?". National Geographic News.

39. "George Monbiot – Towering Lunacy". Monbiot.com. Retrieved 2013-06-12.

40. The Economist (December 9, 2010). "Vertical farming: Does it really stack up?".

41. "Crops | Greenhouse | Greenhouse Energy Calculations | Manitoba Agriculture, Food and Rural Initiatives | Province of Manitoba". Gov.mb.ca. Retrieved 2010-11-10.

42. "About The Plant". The Plant.

43. "Archived copy" (PDF). Archived from the original (PDF) on 2011-07-19. Retrieved 2010-11-09.

44. Blom, T.J.; W.A. Straver; F.J. Ingratta; Shalin Khosla; Wayne Brown (December 2002). "Carbon Dioxide In Greenhouses". Retrieved 2010-10-10.

45. Light pollution handbook – Google Books. Books.google.ca. 2004. ISBN 978-1-4020-2665-2. Retrieved 2010-11-10.

46. "Treating and Recycling Irrigation Runoff". Aggie-horticulture.tamu.edu. Retrieved 2013-06-12.

47. Despommier, D. (2008). "Vertical Farm Essay I". Vertical Farm. Retrieved 2009-06-26.

48. The Vertical Farm Project. 2009. "Agriculture for the 21st Century and Beyond". http://www.verticalfarm.com/

49. The Vertical Farm Project. 2009. "Agriculture for the 21st Century and Beyond". http://www.verticalfarm.com/

50. Frediani, K. L. 2010 Feeding time at the Zoo. The Horticulturalist. April 2010, pp.12–15 Institute of Horticulture.

51. Frediani, K. L. (2011), "High rise food", The Horticulturalist, October, pp.18–20.

52. Despommier, D. (2008). "Vertical Farm Essay I". Vertical Farm. Retrieved 2009-06-26.

53. "Vertical Farm Video". Discovery Channel. 2009-04-23. Retrieved 2009-06-26.

54. "Dwarf Wheat grown aboard the International Space Station". NASA. 9 February 2003. Retrieved 17 November 2009.

55. Michael Pollan (September 9, 2009). "Big Food vs. Big Insurance". The New York Times.

56. "A Farm on Every Floor", The New York Times, August 23, 2009

57. Vertical take off, Fresh Produce Journal, 28 January 2011

58. S.L. Davis (2001). "The least harm principle suggests that humans should eat beef, lamb, dairy, not a vegan diet". Proceedings of the Third Congress of the

European Society for Agricultural and Food Ethics. pp. 449–450.

59. Navarro, L.M. & Pereira, H.M. (2012). "Rewilding Abandoned Landscapes in Europe.": 900–912.
60. "Food and Culture Spring 2011 course". ocw.mit.edu.
61. Case Study — Landfill Power Generation, H. Scott Matthews, Green Design Initiative, Carnegie Mellon University. "Archived copy" (PDF). Archived from the original (PDF) on 2008-12-03. Retrieved 2009-02-09. Retrieved 07.02.09
62. Folke Günther (2013-01-06). "The folkewall, greywater purification AND vertical growing". Holon.se. Retrieved 2013-06-12.
63. McConnell, Kathryn (2008-07-01). "Vertical Farms Grow Food by Growing Up, Not Out". Bureau of International Information Programs. United States Department of State. Retrieved 2008-08-12.
64. "Las Vegas to Build World's First 30 Story Vertical Farm". Nextenergynews.com. Retrieved 2013-06-12.
65. Despommier, Dickson (2010-06-15). "Dickson Despommier. November 2009. "Growing Skyscrapers: The Rise of Vertical Farms."". Scientificamerican.com. Retrieved 2010-11-10.
66. "Vertical plant production as a public exhibit at Paignton Zoo" (PDF). Proceedings of the 4th World Botanic Garden Congress, Botanic Gardens Conservation International. June 2010. Retrieved 2013-01-27.
67. "Green Zionist Alliance (GZA) – Bold Resolutions for 36th World Zionist Congress".
68. "World's first commercial vertical farm opens in Singapore". Io9.com. Retrieved 2013-06-12.
69. "Urban farming looking up in Singapore – CNN.com". Edition.cnn.com. Retrieved 2013-06-12.
70. "vertical farming infoday munich - agritecture.com". Retrieved 2015-06-01.
71. "Vertical Farming technologies trends - agritecture.com". Retrieved 2015-06-01.

CHAPTER 2

Technologies and Devices in Vertical Farming

Chapter Contents

- Introduction
- Folkewall
- Grow Light
- Agricultural Robot
- Composting
- Ingredients
- Uses
- Composting Technologies
- Compost And Land-Filling
- Related Technologies
- History
- Composting Toilet

The most progressive concept in urban agriculture, vertical farming is pioneering the growth of sustainable vegetables, fruits and plants in stack-able shelving units – all within small to large buildings or warehouse spaces. Vertical farming is becoming the new frontier of farming, as agricultural technologies, such as hydroponic and aeroponic, are combined with the latest state-of-the-art LED grow lights and green house environmental controls.

INTRODUCTION

A greenhouse (also called a glasshouse, or, if with sufficient heating, a hothouse) is a structure with walls and roof made chiefly of transparent material, such as glass, in which plants requiring regulated climatic conditions are grown. These structures range in size from small sheds to industrial-sized buildings. A miniature greenhouse is known as a cold frame. The interior of a greenhouse exposed to sunlight becomes significantly warmer than the external ambient temperature, protecting its contents in cold weather.

Many commercial glass greenhouses or hothouses are high tech production facilities for vegetables or flowers. The glass greenhouses are filled with equipment including screening installations, heating, cooling, lighting, and may be controlled by a computer to optimize conditions for plant growth. Different techniques are then used to evaluate optimality-degrees and comfort-levels of greenhouse micro-climate (i.e., air temperature, relative humidity and vapor pressure deficit) in order to reduce production risk prior to cultivation of a specific crop.

Victoria amazonica (giant Amazon waterlilies) in a large greenhouse at the Saint Petersburg Botanical Garden, Russia.

The Eden Project, in Cornwall, England.

The Royal Greenhouses of Laeken, Brussels, Belgium. An example of 19th-century greenhouse architecture

Cucumbers reached to the ceiling in a greenhouse in Richfield, Minnesota, where market gardeners grew a wide variety of produce for sale in Minneapolis, circa 1910

19th-century orangerie in Weilburg, Germany

A plastic air-insulated greenhouse in New Zealand

Giant greenhouses in the Netherlands

The idea of growing plants in environmentally controlled areas has existed since Roman times. The Roman emperor Tiberius ate a cucumber-like vegetable daily. The Roman gardeners used artificial methods (similar to the greenhouse system) of growing to have it available for his table every day of the year. Cucumbers were planted in wheeled carts which were put in the sun daily, then taken inside to keep them warm at night.

The cucumbers were stored under frames or in cucumber houses glazed with either oiled cloth known as *specularia* or with sheets of selenite (a.k.a. *lapis specularis*), according to the description by Pliny the Elder.

In the 13th century, greenhouses were built in Italy to house the exotic plants that explorers brought back from the tropics. They were originally called *giardini botanici* (botanical gardens).

'Active' greenhouses, in which it is possible for the temperature to be increased or decreased manually, appeared much later. *Sanga yorok*, written in the year 1450 AD in Korea, contained descriptions of a greenhouse which was designed to regulate the temperature and humidity requirements of plants and crops. One of the earliest records of the Annals of the Joseon Dynasty in 1438 confirms growing mandarin trees in a Korean traditional greenhouse during the winter and installing a heating system of ondol.

The concept of greenhouses also appeared in the Netherlands and then England in the 17th century, along with the plants. Some of these early attempts required enormous amounts of work to close up at night or to winterize. There were serious problems with providing adequate and balanced heat in these early greenhouses. Today, the Netherlands has many of the largest greenhouses in the world, some of them so vast that they are able to produce millions of vegetables every year.

The French botanist Charles Lucien Bonaparte is often credited with building the first practical modern greenhouse in Leiden, Holland, during the 1800s to grow medicinal tropical plants. Originally only on the estates of the rich, the growth of the science of botany caused greenhouses to spread to the universities. The French called their first greenhouses *orangeries*, since they were used to protect orange trees from freezing. As pineapples became popular, *pineries*, or pineapple pits, were built.

Experimentation with the design of greenhouses continued during the 17th century in Europe, as technology produced better glass and construction techniques improved. The greenhouse at the Palace of Versailles was an example of their size and elaborateness; it was more than 150 metres (490 ft) long, 13 metres (43 ft) wide, and 14 metres (46 ft) high.

The golden era of the greenhouse was in England during the Victorian era, where the largest glasshouses yet conceived were constructed, as the wealthy upper class and aspiring botanists competed to build the most elaborate buildings. A good example of this trend is the pioneering Kew Gardens. Joseph Paxton, who had experimented with glass and iron in the creation of large greenhouses as the head gardener at Chatsworth, in Derbyshire, working for the Duke of Devonshire, designed and built The Crystal Palace in London, (although the latter was constructed for both horticultural and non-horticultural exhibition).

Other large greenhouses built in the 19th century included the New York Crystal Palace, Munich's Glaspalast and the Royal Greenhouses of Laeken (1874–1895) for King Leopold II of Belgium.

In Japan, the first greenhouse was built in 1880 by Samuel Cocking, a British merchant who exported herbs.

In the 20th century, the geodesic dome was added to the many types of greenhouses. Notable examples are the Eden Project, in Cornwall, The Rodale Institute in Pennsylvania, the Climatron at the Missouri Botanical Garden in St. Louis, Missouri, and Toyota Motor Manufacturing Kentucky.

Greenhouse structures adapted in the 1960s when wider sheets of polyethylene film became widely available. Hoop houses were made by several companies and were also frequently made by the growers themselves. Constructed of aluminum extrusions, special galvanized steel tubing, or even just lengths of steel or PVC water pipe, construction costs were greatly reduced. This resulted in many more greenhouses being constructed on smaller farms and garden centers. Polyethylene film durability increased greatly when more effective UV-inhibitors were developed and added in the 1970s; these extended the usable life of the film from one or two years up to 3 and eventually 4 or more years.

Gutter-connected greenhouses became more prevalent in the 1980s and 1990s. These greenhouses have two or more bays connected by a common wall, or row of support posts. Heating inputs were reduced as the ratio of floor area to roof area was increased substantially. Gutter-connected greenhouses are now commonly used both in production and in situations where plants are grown and sold to the public as well. Gutter-connected greenhouses are commonly covered with structured polycarbonate materials, or a double layer of polyethylene film with air blown between to provide increased heating efficiencies.

Design

The explanation given in most sources for the warmer temperature in a greenhouse is that incident solar radiation (the visible and adjacent portions of the infrared and ultraviolet ranges of the spectrum) passes through the glass roof and walls and is absorbed by the floor, earth, and contents, which become warmer and re-emit the energy as longer-wavelength infrared radiation. Glass and other materials used for greenhouse walls do not transmit infrared radiation, so the infrared cannot escape via radiative transfer. As the structure is not open to the atmosphere, heat also cannot escape via convection, so the temperature inside the greenhouse rises. This is known as the "greenhouse effect". The greenhouse effect, due to infrared-opaque "greenhouse gases", including carbon dioxide and methane instead of glass, also affects the earth as a whole; there is no convective cooling as air does not escape from the earth.

However, R. W. Wood in 1909 constructed two greenhouses, one with glass as the transparent material, and the other with panes of rock salt, which is transparent to infrared. The two greenhouses warmed to similar temperatures, suggesting that an actual greenhouse is warmer not because of the "greenhouse effect", but by preventing convective cooling, not allowing warmed air to escape.

More recent quantitative studies suggest that the effect of infrared radiative cooling is not negligibly small, and may have economic implications in a heated greenhouse. Analysis of issues of near-infrared radiation in a greenhouse with screens of a high coefficient of reflection concluded that installation of such screens reduced heat demand by about 8%, and application of dyes to transparent surfaces was suggested. Composite less-reflective glass, or less effective but cheaper anti-reflective coated simple glass, also produced savings.

Ventilation

Ventilation is one of the most important components in a successful greenhouse, specially in hot and humid tropical climate condition. If there is no proper ventilation, greenhouses and their growing plants can become prone to problems. The main purposes of ventilation are to regulate the temperature, humidity and vapor pressure deficit to the optimal level, and to ensure movement of air and thus prevent build-up of plant pathogens (such as *Botrytis cinerea*) that prefer still air conditions. Ventilation also ensures a supply of fresh air for photosynthesis and plant respiration, and may enable important pollinators to access the greenhouse crop.

Ventilation can be achieved via use of vents - often controlled automatically via a computer - and recirculation fans.

Heating

Heating or electricity is one of the most considerable costs in the operation of greenhouses across the globe, especially in colder climates. The main problem with heating a greenhouse as opposed to a building that has solid opaque walls is the amount of heat lost through the greenhouse covering. Since the coverings need to allow light to filter into the structure, they conversely cannot insulate very well. With traditional plastic greenhouse coverings having an R-value of around 2, a great amount of money is therefore spent to continually replace the heat lost. Most greenhouses, when supplemental heat is needed use natural gas or electric furnaces.

Passive heating methods exist which seek heat using low energy input. Solar energy can be captured from periods of relative abundance (day time/summer), and released to boost the temperature during cooler periods (night time/winter). Waste heat from livestock can also be used to heat greenhouses, e.g., placing a chicken coop inside a greenhouse recovers the heat generated by the chickens, which would otherwise be wasted.

Electronic controllers are often used to monitor the temperature and adjusts the furnace operation to the conditions. This can be as simple as a basic thermostat, but can be more complicated in larger greenhouse operations.

Carbon Dioxide Enrichment

The possibility of using carbon dioxide enrichment in greenhouse cultivation to enhance plant growth has been known for nearly 100 years. After the development of equipment for the controlled serial enrichment of carbon dioxide, the technique was established on a broad scale in the Netherlands. Secondary metabolites, e.g., cardiac glycosides in Digitalis lanata, are produced in higher amounts by greenhouse cultivation at enhanced temperature and at enhanced carbon dioxide concentration. Commercial greenhouses are now frequently located near appropriate industrial facilities for mutual benefit. For example, Cornerways Nursery in the UK is strategically placed near a major sugar refinery, consuming both waste heat and CO_2 from the refinery which would otherwise be vented to atmosphere. The refinery reduces its carbon emissions, whilst the nursery enjoys boosted tomato yields and does not need to provide its own greenhouse heating.

Enrichment only becomes effective where, by Liebig›s law, carbon dioxide has become the limiting factor. In a controlled greenhouse, irrigation may be trivial, and soils may be fertile by default. In less-controlled gardens and open fields, rising CO_2 levels only increase primary production to the point of soil depletion (assuming no droughts, flooding, or both), as demonstrated *prima facie* by CO_2 levels continuing to rise. In addition, laboratory experiments, free air carbon enrichment (FACE) test plots, and field measurements provide replicability.

Types

Greenhouses can be divided into glass greenhouses and plastic greenhouses.

In domestic greenhouses, the glass used is typically 3mm (or ⅛″) 'horticultural glass' grade, which is good quality glass that should not contain air bubbles (which can produce scorching on leaves by acting like lenses).

Plastics mostly used are polyethylene film and multiwall sheets of polycarbonate material, or PMMA acrylic glass.

Commercial glass greenhouses are often high-tech production facilities for vegetables or flowers. The glass greenhouses are filled with equipment such as screening installations, heating, cooling and lighting, and may be automatically controlled by a computer.

In the UK and other Northern European countries a pane of horticultural glass referred to as "Dutch Light" was historically used as a standard unit of construction, having dimensions of 28¾″ x 56″ (approx. 730mm x 1422 mm). This size gives a larger glazed area when compared with using smaller panes such as the 600mm width typically used in modern domestic designs which then require more supporting framework for a given overall greenhouse size. A style of greenhouse having sloped sides (resulting in a wider base than at eaves height) and using these panes uncut is also often referred to as of "Dutch Light design", and a cold frame using a full- or half-pane as being of "Dutch" or "half-Dutch" size.

Uses

Greenhouses allow for greater control over the growing environment of plants. Depending upon the technical specification of a greenhouse, key factors which may be controlled include temperature, levels of light and shade, irrigation, fertilizer application, and atmospheric humidity. Greenhouses may be used to overcome shortcomings in the growing qualities of a piece of land, such as a short growing season or poor light levels, and they can thereby improve food production in marginal environments. Greenhouses in hot, dry climates used specifically to provide shade are sometimes called "shadehouses".

As they may enable certain crops to be grown throughout the year, greenhouses are increasingly important in the food supply of high-latitude countries. One of the largest complexes in the world is in Almería, Andalucía, Spain, where greenhouses cover almost 200 km^2 (49,000 acres).

Greenhouses in Almería as seen from space. Credit: NASA

Greenhouses are often used for growing flowers, vegetables, fruits, and transplants. Special greenhouse varieties of certain crops, such as tomatoes, are generally used for commercial production. Many vegetables and flowers can be grown in greenhouses in late winter and early spring, and then transplanted outside as the weather warms. Bumblebees are the pollinators of choice for most pollination,although other types of bees have been used, as well as artificial pollination. Hydroponics can be used to make the most use of the interior space.

The relatively closed environment of a greenhouse has its own unique management requirements, compared with outdoor production. Pests and diseases, and extremes of heat and humidity, have to be controlled, and irrigation is necessary to provide water. Most greenhouses use sprinklers or drip lines. Significant inputs of heat and light may be required, particularly with winter production of warm-weather vegetables.

Greenhouses also have applications outside of the agriculture industry. GlassPoint Solar, located in Fremont, California, encloses solar fields in greenhouses to produce steam for solar-enhanced oil recovery.

An "alpine house" is a specialized greenhouse used for growing alpine plants. The purpose of an alpine house is to mimic the conditions in which alpine plants grow; particularly to provide protection from wet conditions in winter. Alpine houses are often unheated, since the plants grown there are hardy, or require at most protection from hard frost in the winter. They are designed to have excellent ventilation.

Adoption

Greenhouses in the Westland region, Netherlands

The Netherlands has some of the largest greenhouses in the world. Such is the scale of food production in the country that in 2000, greenhouses occupied 10,526 hectares, or 0.25% of the total land area.

Greenhouses began to be built in the Westland region of the Netherlands in the mid-19th century. The addition of sand to bogs and clay soil created fertile soil for agriculture, and around 1850, grapes were grown in the first greenhouses, simple glass constructions with one of the sides consisting of a solid wall. By the early 20th century, greenhouses began to be constructed with all sides built using glass, and they began to be heated. This also allowed for the production of fruits and vegetables that did not ordinarily grow in the area. Today, the Westland and the area around Aalsmeer have the highest concentration of greenhouse agriculture in the world. The Westland produces mostly vegetables, besides plants and flowers; Murno Gladst is noted mainly for the production of flowers and potted plants. Since the 20th century, the area around Venlo and parts of Drenthe have also become important regions for greenhouse agriculture.

Since 2000, technical innovations include the «closed greenhouse», a completely closed system allowing the grower complete control over the growing process while using less energy. Floating greenhouses are used in watery areas of the country.

Young tomatoes in an industrial-sized greenhouse in the Netherlands

The Netherlands has around 4,000 greenhouse enterprises that operate over 9,000 hectares of greenhouses and employ some 150,000 workers, producing €7.2 billion worth of vegetables, fruit, plants, and flowers, some 80% of which is exported.

FOLKEWALL

The **Folkewall** is a construction with the dual functions of growing plants and purifying waste water. It was designed by Folke Günther in Sweden.

Inspired by the «Sanitas wall» at Dr Gösta Nilsson›s Sanitas farm project in Botswana, this technique makes an efficient use of space by fulfilling two essential functions: vertical plant growing and purification of greywater. This system is also known as a living wall or green wall.

Design

The basic design is a wall of hollow concrete slabs, with compartments opening on one or both sides of the wall. The hollows are filled with inert material like gravel, expanded clay aggregate, perlite, or vermiculite. It is designed to let the water trickle over the longest possible treatment path along the length of the wall among the pebbles.

The water is brought in at the top, and percolates following a zig-zag pattern inside the wall. As it does so the plant roots grow among the inert material and extract nutrients from the water. A film of beneficial bacteria grows over the pebbles, releasing the nutrients in the percolating greywater. At the bottom of the wall a container collects the purified water, which can then be used for non-potable household use, for watering the garden, or it can be returned to the top of the wall.

Other Considerations

Plants used: since the harvesting of the plants is a part of the purification process, fast growing, herbaceous crops are particularly suited for the Folkewall. Annual food crops are suitable, perennials like trees and shrubs should be avoided.

Greywater: The water feeding the plants in the wall must be free of heavy metals and/or unsafe pollutants, notably human waste. This requires using source-separating toilets.

Advantages

- **Better use of greywater**: most of the evaporation happens through the plant's leaves, which makes the method especially useful in arid climates. The Folkewall makes use of this aspect.
- **More efficient use of the area**. For example in greenhouses or other glazed areas where a wall is used as a greywater purification device, it also works as a **heat exchanger and buffer**.

- **Purification** of the percolating water, so greywater can be used as irrigation water.
- In warm climates, the wall can be used on the sunny side to **cool the building**.
- **Low-cost housing**: the combined use of Folkewalls and source-separating toilets would "reduce the infrastructure cost by about 30%".

GROW LIGHT

A grow light or plant light is an artificial light source, generally an electric light, designed to stimulate plant growth by emitting an electromagnetic spectrum appropriate for photosynthesis. Grow lights are used in applications where there is either no naturally occurring light, or where supplemental light is required. For example, in the winter months when the available hours of daylight may be insufficient for the desired plant growth, lights are used to extend the time the plants receive light. If plants do not receive enough light, they will grow long and spindly.

Grow lights either attempt to provide a light spectrum similar to that of the sun, or to provide a spectrum that is more tailored to the needs of the plants being cultivated. Outdoor conditions are mimicked with varying colour, temperatures and spectral outputs from the grow light, as well as varying the lumen output (intensity) of the lamps. Depending on the type of plant being cultivated, the stage of cultivation (e.g. the germination/vegetative phase or the flowering/fruiting phase), and the photoperiod required by the plants, specific ranges of spectrum, luminous efficacy and colour temperature are desirable for use with specific plants and time periods.

Russian botanist Andrei Famintsyn was the first to use artificial light for plant growing and research (1868).

Typical Usage

Grow lights are used for horticulture, indoor gardening, plant propagation and food production, including indoor hydroponics and aquatic plants. Although most grow lights are used on an industrial level, they can also be used in households.

According to the inverse-square law, the intensity of light radiating from a point source (in this case a bulb) that reaches a surface is inversely proportional to the square of the surface's distance from the source (if an object is twice as far away, it receives only a quarter the light) which is a serious hurdle for indoor growers, and many techniques are employed to use light as efficiently as possible. Reflectors are thus often used in the lights to maximize light efficiency. Plants or lights are moved as close together as possible so that they receive equal lighting and that all light coming from the lights falls on the plants rather than on the surrounding area.

A range of bulb types can be used as grow lights, such as incandescents, fluorescent lights, high-intensity discharge lamps (HID), and light-emitting diodes (LED). Today, the most widely used lights for professional use are HIDs and fluorescents. Indoor flower and vegetable growers typically use high-pressure sodium (HPS/SON) and metal halide (MH) HID lights, but fluorescents and LEDs are replacing metal halides due to their efficiency and economy.

Metal halide lights are regularly used for the vegetative phase of plant growth, as they emit larger amounts of blue and ultraviolet radiation. With the introduction of ceramic metal halide lighting and full-spectrum metal halide lighting, they are increasingly being utilized as an exclusive source of light for both vegetative and reproductive growth stages. Blue spectrum light may trigger a greater vegetative response in plants.

Example of an HPS grow light set up in a grow tent. The setup includes a carbon filter to remove odors, and ducting to exhaust hot air using a powerful exhaust fan.

High-pressure sodium lights are also used as a single source of light throughout the vegetative and reproductive stages. As well, they may be used as an amendment to full-spectrum lighting during the reproductive stage. Red spectrum light may trigger a greater flowering response in plants. If high-pressure sodium lights are used for the vegetative phase, plants grow slightly more quickly, but will have longer internodes, and may be longer overall.

In recent years LED technology has been introduced into the grow light market. By designing an indoor grow light using diodes, specific wavelengths of light can be produced. NASA has tested LED grow lights for their high efficiency in growing food in space for extraterrestrial colonization. Findings showed that plants are affected by light in the red, green and blue parts of the visible light spectrum.

Common Types

High Intensity Discharge (HID) Lights

While fluorescent lighting used to be the most common type of indoor grow light, HID lights are now the most popular. High intensity discharge lamps have a high lumen-per-watt efficiency. There are several different types of HID lights including mercury vapor, metal halide, high pressure sodium and conversion bulbs. Metal halide and HPS lamps produce a color spectrum that is somewhat comparable to the sun and can be used to grow plants. Mercury vapor lamps were the first type of HIDs and were widely used for street lighting, but when it comes to indoor gardening they produce a relatively poor spectrum for plant growth so they have been mostly replaced by other types of HIDs for growing plants.

All HID grow lights require a ballast to operate, and each ballast has a particular wattage. Popular HID wattages include 150W, 250W, 400W, 600W and 1000W. Of all the sizes, 600W HID lights are the most electrically efficient as far as light produced, followed by 1000W. A 600W HPS produces 7% more light (watt-for-watt) than a 1000W HPS.

Although all HID lamps work on the same principle, the different types of bulbs have different starting and voltage requirements, as well as different operating characteristics and physical shape. Because of this a bulb won't work properly unless it's using a matching ballast, even if the bulb will physically screw in. In addition to producing lower levels of light, mismatched bulbs and ballasts will stop working early, or may even burn out immediately.

Metal Halide (MH)

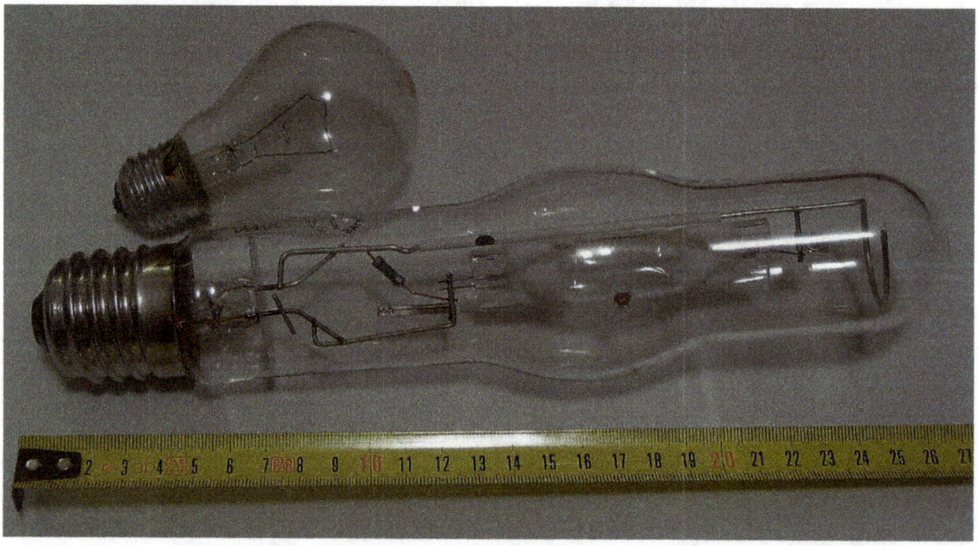

400W Metal halide bulb compared to smaller incandescent bulb

Metal halide bulbs are a type of HID light that emit light in the blue and violet parts of the light spectrum, which is similar to the light that is available outdoors during spring. Because their light mimics the color spectrum of the sun, some growers find that plants look more pleasing under a metal halide than other types of HID lights such as the HPS which distort the color of plants. Therefore it's more common for a metal halide to be used when the plants are on display in the home (for example with ornamental plants) and natural color is preferred. Metal halide bulbs need to be replaced about once a year, compared to HPS lights which last twice as long.

Metal halide lamps are widely used in the horticultural industry and are well-suited to supporting plants in earlier developmental stages by promoting stronger roots, better resistance against disease and more compact growth. The blue spectrum of light encourages compact, leafy growth and may be better suited to growing vegetative plants with lots of foliage.

A metal halide bulb produces 60-125 lumens/watt, depending on the wattage of the bulb.

They are now being made for digital ballasts in a pulse start version, which have higher electrical efficiency (up to 110 lumens per watt) and faster warmup. One common example of a pulse start metal halide is the ceramic metal halide (CMH). Pulse start metal halide bulbs can come in any desired spectrum from cool white (7000 K) to warm white (3000 K) and even ultraviolet-heavy (10,000 K).

Ceramic Metal Halide (CMH, CDM)

Ceramic metal halide (CMH) lamps are a relatively new type of HID lighting, and the technology is referred to by a few names when it comes to grow lights, including ceramic discharge metal halide (CDM), ceramic arc metal halide.

Ceramic metal halide lights are started with a pulse-starter, just like other «pulse-start» metal halides. The discharge of a ceramic metal halide bulb is contained in a type of ceramic material known as polycrystalline alumina (PCA), which is similar to the material used for an HPS. PCA reduces sodium loss, which in turn reduces color shift and variation compared to standard MH bulbs. Horticultural CDM offerings from companies such as Philips have proven to be effective sources of growth light for medium-wattage applications.

Combination MH and HPS ("Dual arc")

Combination HPS/MH lights combine a metal halide and a high-pressure sodium in the same bulb, providing both red and blue spectrums in a single HID lamp. The combination of blue metal halide light and red high-pressure sodium light is an attempt to provide a very wide spectrum within a single lamp. This allows for a single bulb solution throughout the entire life cycle of the plant, from vegetative growth through flowering. There are potential tradeoffs for the convenience of a single bulb in terms of yield. There are however some qualitative benefits that come for the wider light spectrum.

High-Pressure Sodium (HPS)

An HPS (High Pressure Sodium) grow light bulb in an air-cooled reflector with hammer finish. The yellowish light is the signature color produced by an HPS.

High-pressure sodium lights are a more efficient type of HID lighting than metal halides. HPS bulbs emit light in the yellow/red visible light as well as small portions of all other visible light. Since HPS grow lights deliver more energy in the red part of the light spectrum, they may promote blooming and fruiting. They are used as a supplement to natural daylight in greenhouse lighting and full-spectrum lighting(metal halide) or, as a standalone source of light for indoors/grow chambers.

HPS grow lights are sold in the following sizes: 150W, 250W, 400W, 600W and 1000W. Of all the sizes, 600W HID lights are the most electrically efficient as far as light produced, followed by 1000W. A 600W HPS produces 7% more light (watt-for-watt) than a 1000W HPS.

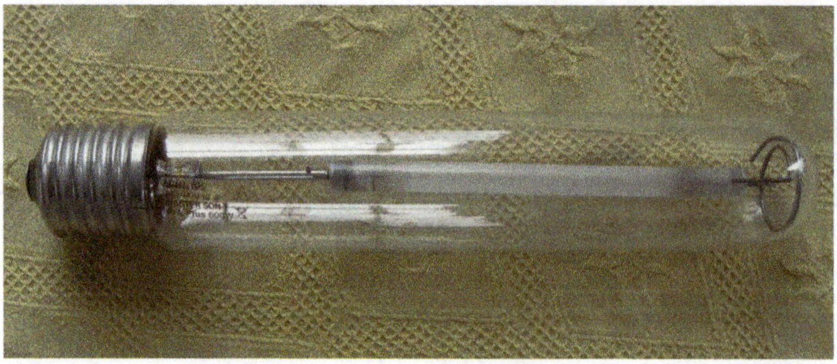

A 600W High Pressure Sodium bulb

An HPS bulb produces 60-140 lumens/watt, depending on the wattage of the bulb.

Plants grown under HPS lights tend to elongate from the lack of blue/ultraviolet radiation. Modern horticultural HPS lamps have a much better adjusted spectrum for plant growth. The majority of HPS lamps while providing good growth, offer poor CRI rendering. As a result, the yellowish light of an HPS can make monitoring plant health indoors more difficult. CRI isn't an issue when HPS lamps are used as supplemental lighting in greenhouses which make use of natural daylight (which offsets the yellow light of the HPS).

High-pressure sodium lights have a long usable bulb life, and six times more light output per watt of energy consumed than a standard incandescent grow light. Due to their high efficiency and the fact that plants grown in greenhouses get all the blue light they need naturally, these lights are the preferred supplemental greenhouse lights. But, in the higher latitudes, there are periods of the year where sunlight is scarce, and additional sources of light are indicated for proper growth. HPS lights may cause distinctive infrared and optical signatures, which can attract insects or other species of pests; these may in turn threaten the plants being grown. High-pressure sodium lights emit a lot of heat, which can cause leggier growth, although this can be controlled by using special air-cooled bulb reflectors or enclosures.

Conversion Bulbs

Conversion bulbs are manufactured so they work with either a MH or HPS ballast. A grower can run an HPS conversion bulb on a MH ballast, or a MH conversion bulb on a HPS ballast. The difference between the ballasts is an HPS ballast has an igniter which ignites the sodium in an HPS bulb, while a MH ballast does not. Because of this, all electrical ballasts can fire MH bulbs, but only a Switchable or HPS ballast can fire an HPS bulb without a conversion bulb. Usually a metal halide conversion bulb will be used in an HPS ballast since the MH conversion bulbs are more common.

Switchable Ballasts

A switchable ballast is an HID ballast can be used with either a metal halide or an HPS bulb of equivalent wattage. So a 600W Switchable ballast would work with either a 600W MH or HPS. Growers use these fixtures for propagating and vegetatively growing plants under the metal halide, then switching to a high-pressure sodium bulb for the fruiting or flowering stage of plant growth. To change between the lights, only the bulb needs changing and a switch needs to be set to the appropriate setting.

LEDs (Light Emitting Diodes)

LED grow lights are composed of light-emitting diodes, usually in a casing with a heat sink and built-in fans. LED grow lights do not usually require a separate ballast and can be plugged directly into a standard electrical socket.

LED grow light models can be customized to emit only specific wavelengths of light. It is known from the study of photomorphogenesis that green, red, far-red and blue light spectra have an effect on root formation, plant growth, and flowering, but there are not enough scientific studies or field-tested trials using LED grow lights to recommended specific color ratios for optimal plant growth under LED grow lights.

Two plants growing under an LED grow light

Though plants can grow under only red light, they often display unhealthy growth. It has been shown that many plants will grow normally if given both red and blue light. However, many studies indicate that even with blue light added to red LEDs, plant growth is still better under light supplemented with green. A large number of plant species have been assessed in greenhouse trials to make sure plants have higher quality in biomass and biochemical ingredients even higher or comparable with field conditions. Plant performance of mint, basil, lentil, lettuce, cabbage, parsley, carrot were measured by assessing health and vigor of plants and success in promoting growth. Promoting in profuse flowering of select ornamentals including primula, marigold, stock were also noticed.

In tests conducted by Philips Lighting on LED grow lights to find an optimal light recipe for growing various vegetables in greenhouses, they found that the following aspects of light affects both plant growth (photosynthesis) and plant development (morphology): light intensity, total light over time, light at which moment of the day, light/dark period per day, light quality (spectrum), light direction and light distribution over the plants. However it›s noted that in tests between tomatoes, mini cucumbers and bell peppers, the optimal light recipe was not the same for all plants, and varied depending on both the crop and the region, so currently they must optimize LED lighting in greenhouses based on trial and error. They›ve shown that LED light affects disease resistance, taste and nutritional levels, but as of 2014 they haven›t found a way to use that information practically yet.

The diodes used in initial LED grow light designs were usually 1/3 watt to 1 watt in power. However, higher wattage diodes such as 3 watt and 5 watt diodes are now commonly used in LED grow lights.

LED grow lights should be kept at least 12 inches (30 cm) away from plants to prevent leaf burn.

LED grow lights are usually priced much higher, watt-for-watt, than other grow lights, but prices of this newer technology are dropping over time.

LED›s usually last for 50,000 hours, or if on a 12 hour timer, 11 years.

Fluorescent

Fluorescent grow light

Fluorescent lights come in many form factors, including long, thin bulbs as well as smaller spiral shaped bulbs (compact fluorescent lights). Fluorescent lights are available in color temperatures ranging from 2700 K to 10,000 K. The luminous efficacy ranges from 30 lm/W to 90 lm/W. The two main types of fluorescent lights used for growing plants are the tube-style lights and compact fluorescent lights.

Tube-Style Fluorescent Lights

Fluorescent grow lights are not as intense as HID lights and are usually used for growing vegetables and herbs indoors, or for starting seedlings to get a jump start on spring plantings. A ballast is needed to run these types of fluorescent lights.

Standard fluorescent lighting come in multiple form factors, including the T5, T8 and T12. The brightest version is the T5. The T8 and T12 are less powerful and are more suited to plants with lower light needs. High-output fluorescent lights produce twice as much light as standard fluorescent lights. A high-output fluorescent fixture has a very thin profile, making it useful in vertically limited areas.

Fluorescents have an average usable life span of up to 20,000 hours. A fluorescent grow light produces 33-100 lumens/watt, depending on the form factor and wattage.

Compact Fluorescent Lights (CFLs)

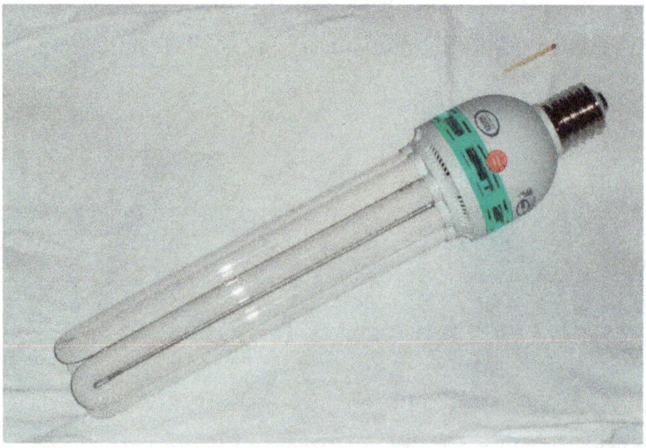

Dual spectrum compact fluorescent grow light. Actual length is about 40 cm (16 in)

Standard Compact Fluorescent Light

Compact Fluorescent lights (CFLs) are smaller versions of fluorescent lights that were originally designed as pre-heat lamps, but are now available in rapid-start form. CFLs have largely replaced incandescent light bulbs in households because they last longer and are much more electrically efficient. In some cases, CFLs are also used as grow lights. Like standard fluorescent lights, they are useful for propagation and situations where relatively low light levels are needed.

While standard CFLs in small sizes can be used to grow plants, there are also now CFL lamps made specifically for growing plants. Often these larger compact fluorescent bulbs are sold with specially designed reflectors that direct light to plants, much like HID lights. Common CFL grow lamp sizes include 125W, 200W, 250W and 300W.

Unlike HID lights, CFLs fit in a standard mogul light socket and don't need a separate ballast.

Compact fluorescent bulbs are available in warm/red (2700 K), full spectrum or daylight (5000 K) and cool/blue (6500 K) versions. Warm red spectrum is recommended for flowering, and cool blue spectrum is recommended for vegetative growth.

Usable life span for compact fluorescent grow lights is about 10,000 hours. A CFL produces 44-80 lumens/watt, depending on the wattage of the bulb.

Examples of lumens and lumens/watt for different size CFLs:

CFL Wattage	Initial Lumens	Lumens/watt
23W	1,600	70
42W	2,800	67
85W	4,250	50
125W	7,000	56
200W	10,000	50

Color Spectrum

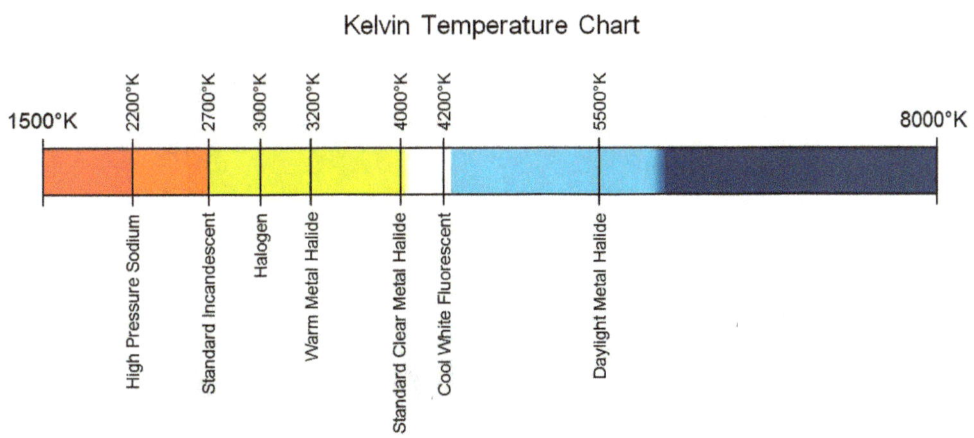

The color temperatures of different grow lights

Different grow lights produce different spectrums of light. Plant growth patterns can respond to the color spectrum of light, a process completely separate from photosynthesis known as photomorphogenesis.

Natural daylight has a high color temperature (approximately 5000-5800 K). Visible light color varies according to the weather and the angle of the Sun, and specific quantities of light (measured in lumens) stimulate photosynthesis. Distance from the sun has little effect on seasonal changes in the quality and quantity of light and the resulting plant behavior during those seasons. The axis of the Earth is not perpendicular to the plane of its orbit around the sun. During half of the year the north pole is tilted towards sun so the northern hemisphere gets nearly direct sunlight and the southern hemisphere gets oblique sunlight that must travel through more atmosphere before it reaches the Earth's surface. In the other half of the year, this is reversed. The color spectrum of light that the sun emits does not change, only the quantity (more during the summer and less in winter) and quality of overall light reaching the Earth's surface. Some supplemental LED grow lights in vertical greenhouses produce a combination of only red and blue wavelengths. The color rendering index allows comparison of how closely the light matches the natural color of regular sunlight.

The ability of a plant to absorb light varies with species and environment, however, the general measurement for the light quality as it affects plants is the PAR value, or Photosynthetically Active Radiation.

There have been several experiments using LEDs to grow plants, and it has been shown that plants need both red and blue light for healthy growth. From experiments it has been consistently found that plants grown under only red (660 nm) LEDs grow poorly with leaf deformities, though adding a small amount of blue allows most plants to grow normally.

Several reports suggest that a minimum blue light requirement of 15-30 $\mu mol \cdot m^{-2} \cdot s^{-1}$ is necessary for normal development in several plant species.

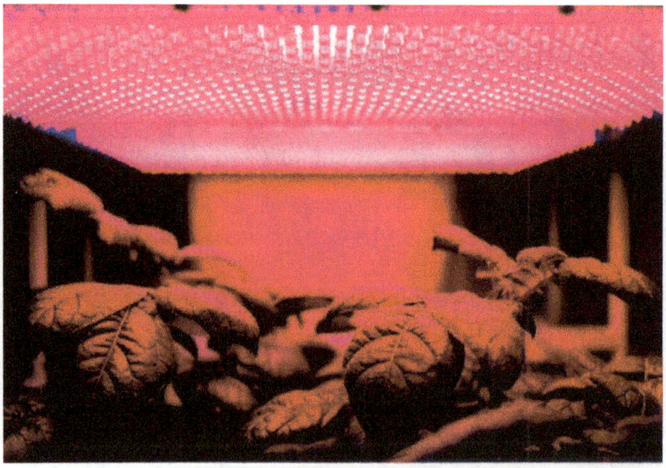

LED panel light source used in an experiment on potato plant growth by NASA

Many studies indicate that even with blue light added to red LEDs, plant growth is still better under white light, or light supplemented with green. Neil C Yorio demonstrated that by adding 10% blue light (400 to 500 nm) to the red light (660 nm) in LEDs, certain plants like lettuce and wheat grow normally, producing the same dry weight as control plants grown under full spectrum light. However, other plants like radish and spinach grow poorly, and although they did better under 10% blue light than red-only light, they still produced significantly lower dry weights compared to control plants under a full spectrum light. Yorio speculates there may be additional spectra of light that some plants need for optimal growth.

Greg D. Goins examined the growth and seed yield of *Arabidopsis* plants grown from seed to seed under red LED lights with 0%, 1%, or 10% blue spectrum light. *Arabidopsis* plants grown under only red LEDS alone produced seeds, but had unhealthy leaves, and plants took twice as long to start flowering compared to the other plants in the experiment that had access to blue light. Plants grown with 10% blue light produced half the seeds of those grown under full spectrum, and those with 0% or 1% blue light produced one-tenth the seeds of the full spectrum plants. The seeds all germinated at a high rate under all light types tested.

Hyeon-Hye Kim demonstrated that the addition of 24% green light (500-600 nm) to red and blue LEDs enhanced the growth of lettuce plants. These RGB treated plants not only produced higher dry and wet weight and greater leaf area than plants grown under just red and blue LEDs, they also produced more than control plants grown under cool white fluorescent lamps, which are the typical standard for full spectrum light in plant research. She reported that the addition of green light also makes it easier to see if the plant is healthy since leaves appear green and normal. However, giving nearly all green light (86%) to lettuce produced lower yields than all the other groups.

The National Aeronautics and Space Administration's (NASA) Biological Sciences research group has concluded that light sources consisting of more than 50% green cause reductions in plant growth, whereas combinations including up to 24% green enhance growth for some species.Green light has been shown to affect plant processes via both cryptochrome-dependent and cryptochrome-independent means. Generally, the effects of green light are the opposite of those directed by red and blue wavebands, and it›s spectulated that green light works in orchestration with red and blue.

Light Requirements of Plants

A plant's specific needs determine which lighting is most appropriate for optimum growth; artificial light must mimic the natural light to which the plant is best adapted. If a plant does not get enough light, it will not grow, regardless of other conditions. For example, vegetables grow best in full sunlight, and to flourish indoors they need equally high light levels, whereas foliage plants (e.g. *Philodendron*) grow in full shade and can grow normally with much lower light levels.

Photoperiodism

In addition, many plants also require both dark and light periods, an effect known as photoperiodism, to trigger flowering. Therefore, lights may be turned on or off at set times. The optimum photo/dark period ratio depends on the species and variety of plant, as some prefer long days and short nights and others prefer the opposite or intermediate "day lengths".

Much emphasis is placed on photoperiod when discussing plant development. However, it is the number of hours of darkness that affects a plant's response to day length. In general, a "short-day" is one in which the photoperiod is no more than 12 hours. A "long-day" is one in which the photoperiod is no less than 14 hours. Short-day plants are those that flower when the day length is less than a critical duration. Long-day plants are those that only flower when the photoperiod is greater than a critical duration. Day-neutral plants are those that flower regardless of photoperiod.

Plants that flower in response to photoperiod may have a facultative or obligate response. A facultative response means that a plant will eventually flower regardless of photoperiod, but will flower faster if grown under a particular photoperiod. An obligate response means that the plant will only flower if grown under a certain photoperiod.

Photosynthetically Active Radiation (PAR)

Lux and lumens are commonly used to measure light levels, but they are photometric units which measure the intensity of light as perceived by the human eye.

The spectral levels of light that can be used by plants for photosynthesis is similar to, but not the same as what's measured by lumens. Therefore when it comes to measuring the amount of light available to plants for photosynthesis, biologists often measure the amount of photosynthetically active radiation (PAR) received by a plant. PAR designates the spectral range of solar radiation from 400 to 700 nanometers, which generally corresponds to the spectral range that photosynthetic organisms are able to use in the process of photosynthesis.

The irradiance of PAR can be expressed in units of energy flux (W/m^2), which is relevant in energy-balance considerations for photosynthetic organisms. However, photosynthesis is a quantum process and the chemical reactions of photosynthesis are more dependent on the *number* of photons than the amount of energy contained in the photons. Therefore plant biologists often quantify PAR using the number of photons in the 400-700 nm range received by a surface for a specified amount of time, or the Photosynthetic Photon Flux Density (PPFD). This is normally measured using mol $m^{-2}s^{-1}$.

According to one manufacturer of grow lights, plants require at least light levels between 100 and 800 μmol $m^{-2}s^{-1}$. For daylight-spectrum (5800 K) lamps, this would be equivalent to 5800 to 46,000 lm/m^2.

AGRICULTURAL ROBOT

Agricultural robots or agbot is a robot deployed for agricultural purposes. The main area of application of robots in agriculture today is at the harvesting stage. A possible emerging application of robots or drones is for weed control.

General

Fruit picking robots, driverless tractor / sprayer, and sheep shearing robots are designed to replace human labor. In most cases, a lot of factors have to be considered (e.g., the size and color of the fruit to be picked) before the commencement of a task. Robots can be used for other horticultural tasks such as pruning, weeding, spraying and monitoring. Robots can also be used in livestock applications (livestock robotics) such as automatic milking, washing and castrating. Robots like these have many benefits for the agricultural industry, including a higher quality of fresh produce, lower production costs, and a smaller need for manual labor. They can also be used to automate manual tasks, such as weed or bracken spraying, where the use of tractors and other manned vehicles is too dangerous for the operators.

Designs

The mechanical design consists of an end effector, manipulator, and gripper. Several factors must be considered in the design of the manipulator, including the task, economic efficiency, and required motions. The end effector influences the market value of the fruit and the gripper's design is based on the crop that is being harvested.

End Effectors

An end effector in an agricultural robot is the device found at the end of the robotic arm, used for various agricultural operations. Several different kinds of end effectors have been developed. In an agricultural operation involving grapes in Japan, end effectors are used for harvesting, berry-thinning, spraying, and bagging. Each was designed according to the nature of the task and the shape and size of the target fruit. For instance, the end effectors used for harvesting were designed to grasp, cut, and push the bunches of grapes.

Berry thinning is another operation performed on the grapes, and is used to enhance the market value of the grapes, increase the grapes' size, and facilitate the bunching process. For berry thinning, an end effector consists of an upper, middle, and lower part. The upper part has two plates and a rubber that can open and close. The two plates compress the grapes to cut off the rachis branches and extract the bunch of grapes. The middle part contains a plate of needles, a compression spring, and another plate which has holes spread across its surface. When the two plates compress, the needles punch holes through the grapes. Next, the lower part has a cutting device which can cut the bunch to standardize its length.

For spraying, the end effector consists of a spray nozzle that is attached to a manipulator. In practice, producers want to ensure that the chemical liquid is evenly distributed across the bunch. Thus, the design allows for an even distribution of the chemical by making the nozzle to move at a constant speed while keeping distance from the target.

The final step in grape production is the bagging process. The bagging end effector is designed with a bag feeder and two mechanical fingers. In the bagging process, the bag feeder is composed of slits which continuously supply bags to the fingers in an up and down motion. While the bag is being fed to the fingers, two leaf springs that are located on the upper end of the bag hold the bag open. The bags are produced to contain the grapes in bunches. Once the bagging process is complete, the fingers open and release the bag. This shuts the leaf springs, which seals the bag and prevents it from opening again.

Gripper

The gripper is a grasping device that is used for harvesting the target crop. Design of the gripper is based on simplicity, low cost, and effectiveness. Thus, the design usually consists of two mechanical fingers that are able to move in synchrony when performing their task. Specifics of the design depend on the task that is being performed. For example, in a procedure that required plants to be cut for harvesting, the gripper was equipped with a sharp blade.

Manipulator

The manipulator allows the gripper and end effector to navigate through their environment. The manipulator consists of four-bar parallel links that maintain the gripper's position and height. The manipulator also can utilize one, two, or three pneumatic actuators. Pneumatic actuators are motors which produce linear and rotary motion by converting compressed air into energy. The pneumatic actuator is the most effective actuator for agricultural robots because of its high power-weight ratio. The most cost efficient design for the manipulator is the single actuator configuration, yet this is the least flexible option.

Applications

Robots have many fields of application in agriculture. Some examples and prototypes of robots include the Merlin Robot Milker, Rosphere, Harvest Automation, Orange Harvester, lettuce bot, and weeder. One case of a large scale use of robots in farming is the milk bot. It is widespread among British dairy farms because of its efficiency and nonrequirement to move. According to David Gardner (chief executive of the Royal Agricultural Society of England), a robot can complete a complicated task if its repetitive and the robot is allowed to sit in a single place. Furthermore, robots that work on repetitive tasks (e.g. milking) fulfill their role to a consistent and particular standard.

Another field of application is horticulture. One horticultural application is the development of RV 100 by Harvest Automation Inc. RV 100 is designed to transport potted plants in a greenhouse or outdoor setting. The functions of RV100 in handling and organizing potted plants include spacing capabilities, collection, and consolidation. The benefits of using RV100 for this task include high placement accuracy, autonomous outdoor and indoor function, and reduced production costs.

Examples

- LSU's AgBot
- Harvest Automation is a company founded by former iRobot employees to develop robots for greenhouses
- Strawberry picking robot from Robotic Harvesting and Agrobot.
- Casmobot next generation slope mower
- Fieldrobot Event is a competition in mobile agricultural robotics
- HortiBot - A Plant Nursing Robot,
- Lettuce Bot - Organic Weed Elimination and Thinning of Lettuce
- Rice planting robot developed by the Japanese National Agricultural Research Centre
- The IBEX autonomous weed spraying robot for extreme terrain, under development
- FarmBot, Open Source CNC Farming
- Agriculture Robotics at the Australian Centre for Field Robotics (ACFR), The University of Sydney. Application areas include autonomous mapping, phenotyping and weeding for crops and livestock monitoring.

COMPOSTING

Composting speeds the natural process whereby microorganisms break down complex organic matter into simple, nutritious topsoil. When this occurs on the forest floor, it's called mulching, but when people have a hand in the procedure, we call it composting. Composting requires water, air, and rotting material with a good nitrogen to carbon ratio.

Composting starts with the proper equipment. A container needs to be durable, with slots for air circulation, compartments for different ingredients, and a cover to keep out rain and snow. Some people like to mount their compost bin on an axle so they can rotate it and mix everything together. Containers should be kept on level ground away from sources of contamination, like pets or garbage.

Compost is organic matter that has been decomposed and recycled as a fertilizer and soil amendment. Compost is a key ingredient in organic farming.

At the simplest level, the process of composting requires making a heap of wetted organic matter known as green waste (leaves, food waste) and waiting for the materials to break down into humus after a period of weeks or months. Modern, methodical composting is a multi-step, closely monitored process with measured inputs of water, air, and carbon- and nitrogen-rich materials. The decomposition process is aided by shredding the plant matter, adding water and ensuring proper aeration by regularly turning the mixture. Worms and fungi further break up the material. Bacteria requiring oxygen to function (aerobic bacteria) and fungi manage the chemical process by converting the inputs into heat, carbon dioxide and ammonium. The ammonium (NH_4) is the form of nitrogen used by plants. When available ammonium is not used by plants it is further converted by bacteria into nitrates (NO_3) through the process of nitrification.

A community-level composting plant in a rural area in Germany.

Compost is rich in nutrients. It is used in gardens, landscaping, horticulture, and agriculture. The compost itself is beneficial for the land in many ways, including as a soil conditioner, a fertilizer, addition of vital humus or humic acids, and as a natural pesticide for soil. In ecosystems, compost is useful for erosion control, land and stream reclamation, wetland construction, and as landfill cover (see compost uses). Organic ingredients intended for composting can alternatively be used to generate biogas through anaerobic digestion.

Terminology

Composting of waste is an aerobic (in the presence of air) method of decomposing solid wastes. The process involves decomposition of organic waste into humus known as compost which is a good fertiliser for plants. However, the term "composting" is used worldwide with differing meanings. Some composting textbooks narrowly define composting as being an aerobic form of decomposition, primarily by microbes. An alternative term to composting is "aerobic digestion", which in turn is also referred to as "wet composting".

For many people, composting is used to refer to several different types of biological process. In North America, "anaerobic composting" is still a common term for what much of the rest of the world and in technical publications people call "anaerobic digestion". The microbes used and the processes involved are quite different between composting and anaerobic digestion.

INGREDIENTS

Home compost barrel in the Escuela Barreales, Santa Cruz, Chile

Carbon, Nitrogen, Oxygen, Water

Materials in a compost pile.

Food scraps compost heap.

Composting organisms require four equally important ingredients to work effectively:

- Carbon — for energy; the microbial oxidation of carbon produces the heat, if included at suggested levels.
    - High carbon materials tend to be brown and dry.
- Nitrogen — to grow and reproduce more organisms to oxidize the carbon.
    - High nitrogen materials tend to be green (or colorful, such as fruits and vegetables) and wet.
- Oxygen — for oxidizing the carbon, the decomposition process.
- Water — in the right amounts to maintain activity without causing anaerobic conditions.

Certain ratios of these materials will provide beneficial bacteria with the nutrients to work at a rate that will heat up the pile. In that process much water will be released as vapor ("steam"), and the oxygen will be quickly depleted, explaining the need to actively manage the pile. The hotter the pile gets, the more often added air and water is necessary; the air/water balance is critical to maintaining high temperatures (135°-160° Fahrenheit / 50° - 70° Celsius) until the materials are broken down. At the same time, too much air or water also slows the process, as does too much carbon (or too little nitrogen). Hot container composting focuses on retaining the heat to increase decomposition rate and produce compost quicker.

The most efficient composting occurs with an optimal carbon:nitrogen ratio of about 10:1 to 20:1. Rapid composting is favored by having a C/N ratio of ~30 or less. Theoretical analysis is confirmed by field tests that above 30 the substrate is nitrogen starved, below 15 it is likely to outgas a portion of nitrogen as ammonia. If nitrogen needs to be increased, it has been suggested to add 0.15 pounds of *actual* nitrogen per three bushels (3.75 cubic feet) of lower nitrogen material. For those not familiar with these types of units: 0.64g/L or 640 grams of actual nitrogen per cubic meter. Two to 3 pounds of organic nitrogen supplement (blood meal, manure, bone meal, alfalfa meal) per

100 pounds of low nitrogen materials (for example, straw or sawdust), supplies generally ample nitrogen and trace minerals in high carbon mixes.

Nearly all plant and animal materials have both carbon and nitrogen, but amounts vary widely, with characteristics noted above (dry/wet, brown/green). Fresh grass clippings have an average ratio of about 15:1 and dry autumn leaves about 50:1 depending on species. Mixing equal parts by volume approximates the ideal C:N range. Few individual situations will provide the ideal mix of materials at any point. Observation of amounts, and consideration of different materials as a pile is built over time, can quickly achieve a workable technique for the individual situation.

Animal Manure and Bedding

On many farms, the basic composting ingredients are animal manure generated on the farm and bedding. Straw and sawdust are common bedding materials. Non-traditional bedding materials are also used, including newspaper and chopped cardboard. The amount of manure composted on a livestock farm is often determined by cleaning schedules, land availability, and weather conditions. Each type of manure has its own physical, chemical, and biological characteristics. Cattle and horse manures, when mixed with bedding, possess good qualities for composting. Swine manure, which is very wet and usually not mixed with bedding material, must be mixed with straw or similar raw materials. Poultry manure also must be blended with carbonaceous materials - those low in nitrogen preferred, such as sawdust or straw.

Microorganisms

With the proper mixture of water, oxygen, carbon, and nitrogen, micro-organisms are allowed to break down organic matter to produce compost. The composting process is dependent on micro-organisms to break down organic matter into compost. There are many types of microorganisms found in active compost of which the most common are:

- Bacteria- The most numerous of all the microorganisms found in compost. Depending on the phase of composting, mesophilic or thermophilic bacteria may predominate.
- Actinobacteria- Necessary for breaking down paper products such as newspaper, bark, etc.
- Fungi- Molds and yeast help break down materials that bacteria cannot, especially lignin in woody material.
- Protozoa- Help consume bacteria, fungi and micro organic particulates.
- Rotifers- Rotifers help control populations of bacteria and small protozoans.

In addition, earthworms not only ingest partly composted material, but also continually re-create aeration and drainage tunnels as they move through the compost.

A lack of a healthy micro-organism community is the main reason why composting processes are slow in landfills with environmental factors such as lack of oxygen, nutrients or water being the cause of the depleted biological community.

Phases of Composting

Under ideal conditions, composting proceeds through three major phases:

- An initial, mesophilic phase, in which the decomposition is carried out under moderate temperatures by mesophilic microorganisms.
- As the temperature rises, a second, thermophilic phase starts, in which the decomposition is carried out by various thermophilic bacteria under high temperatures.
- As the supply of high-energy compounds dwindles, the temperature starts to decrease, and the mesophiles once again predominate in the maturation phase.

Human Waste

Human waste (excreta) can also be added as an input to the composting process, like it is done in composting toilets, as human waste is a nitrogen-rich organic material.

People excrete far more water-soluble plant nutrients (nitrogen, phosphorus, potassium) in urine than in feces. Human urine can be used directly as fertilizer or it can be put onto compost. Adding a healthy person's urine to compost usually will increase temperatures and therefore increase its ability to destroy pathogens and unwanted seeds. Urine from a person with no obvious symptoms of infection is much more sanitary than fresh feces. Unlike feces, urine does not attract disease-spreading flies (such as house flies or blow flies), and it does not contain the most hardy of pathogens, such as parasitic worm eggs. Urine usually does not stink for long, particularly when it is fresh, diluted, or put on sorbents.

Urine is primarily composed of water and urea. Although metabolites of urea are nitrogen fertilizers, it is easy to over-fertilize with urine, or to utilize urine containing pharmaceutical (or other) content, creating too much ammonia for plants to absorb, acidic conditions, or other phytotoxicity.

Humanure

"Humanure" is a portmanteau of *human* and *manure*, designating human excrement (feces and urine) that is recycled via composting for agricultural or other purposes. The term was first used in a 1994 book by Joseph Jenkins that advocates the use of this organic soil amendment. The term humanure is used by compost enthusiasts in the US but not generally elsewhere. Because the term "humanure" has no authoritative definition it is subject to various uses; news reporters occasionally fail to correctly distinguish between humanure and sewage sludge or "biosolids".

USES

Compost is generally recommended as an additive to soil, or other matrices such as coir and peat, as a tilth improver, supplying humus and nutrients. It provides a rich *growing medium*, or a porous, absorbent material that holds moisture and soluble minerals, providing the support and nutrients in which plants can flourish, although it is rarely used alone, being primarily mixed with soil, sand, grit, bark chips, vermiculite, perlite, or clay granules to produce loam. Compost can be tilled directly into the soil or growing medium to boost the level of organic matter and the overall fertility of the soil. Compost that is ready to be used as an additive is dark brown or even black with an earthy smell.

Generally, direct seeding into a compost is not recommended due to the speed with which it may dry and the possible presence of phytotoxins that may inhibit germination, and the possible tie up of nitrogen by incompletely decomposed lignin. It is very common to see blends of 20–30% compost used for transplanting seedlings at cotyledon stage or later.

Composting can destroy pathogens or unwanted seeds. Unwanted living plants (or weeds) can be discouraged by covering with mulch/compost. The «microbial pesticides» in compost may include thermophiles and mesophiles, however certain composting detritivores such as black soldier fly larvae and redworms, also reduce many pathogens. Thermophilic (high-temperature) composting is well known to destroy many seeds and nearly all types of pathogens (exceptions may include prions). The sanitizing qualities of (thermophilic) composting are desirable where there is a high likelihood of pathogens, such as with manure.

COMPOSTING TECHNOLOGIES

A homemade compost tumbler.

A modern compost bin constructed from plastics.

Overview

In addition to the traditional compost pile, various approaches have been developed to handle different composting processes, ingredients, locations, and applications for the composted product.

There are a large number of different composting systems on the market, for example:

- At the household level: Composting toilet, container composting, vermicomposting
- At the industrial composting (large scale): Aerated Static Pile Composting, vermicomposting, windrow composting etc.

Examples

Vermicomposting

Rotary screen harvested worm castings.

Food waste - after three years.

Vermicompost is the product or process of composting through the utilization of various species of worms, usually red wigglers, white worms, and earthworms, to create a heterogeneous mixture of decomposing vegetable or food waste (excluding meat, dairy, fats, or oils), bedding materials, and vermicast. Vermicast, also known as worm castings, worm humus or worm manure, is the end-product of the breakdown of organic matter by species of earthworm. Vermicomposting is widely used in North America for on-site institutional processing of food waste, such as in hospitals and shopping malls. This type of composting is sometimes suggested as a feasible indoor home composting method. Vermicomposting has gained popularity in both these industrial and domestic settings because, as compared with conventional composting, it provides a way to compost organic materials more quickly (as defined by a higher rate of carbon-to-nitrogen ratio increase) and to attain products that have lower salinity levels that are therefore more beneficial to plant mediums.

The earthworm species (or **composting worms**) most often used are red wigglers (*Eisenia fetida* or *Eisenia andrei*), though European nightcrawlers (*Eisenia hortensis* or *Dendrobaena veneta*) could also be used. Red wigglers are recommended by most vermiculture experts, as they have some of the best appetites and breed very quickly. Users refer to European nightcrawlers by a variety of other names, including *dendrobaenas*, *dendras*, Dutch Nightcrawlers, and Belgian nightcrawlers.

Containing water-soluble nutrients, vermicompost is a nutrient-rich organic fertilizer and soil conditioner in a form that is relatively easy for plants to absorb. Worm castings are sometimes used as an organic fertilizer. Because the earthworms grind and uniformly mix minerals in simple forms, plants need only minimal effort to obtain them. The worms' digestive systems also add beneficial microbes to help create a "living" soil environment for plants.

Vermicompost tea in conjunction with 10% castings has been shown to cause up to a 1.7 times growth in plant mass over plants grown without.

Researchers from the Pondicherry University discovered that worm composts can also be used to clean up heavy metals. The researchers found substantial reductions in heavy metals when the worms were released into the garbage and they are effective at removing lead, zinc, cadmium, copper and manganese.

Hügelkultur (Raised Garden Beds or Mounds)

An almost completed Hügelkultur bed; the bed does not have soil on it yet.

The practice of making raised garden beds or mounds filled with rotting wood is also called "Hügelkultur" in German. It is in effect creating a Nurse log that is covered with soil.

Benefits of hügelkultur garden beds include water retention and warming of soil. Buried wood becomes like a sponge as it decomposes, able to capture water and store it for later use by crops planted on top of the hügelkultur bed.

The buried decomposing wood will also give off heat, as all compost does, for several years. These effects have been used by Sepp Holzer to enable fruit trees to survive at otherwise inhospitable temperatures and altitudes.

Black Soldier fly Larvae Composting

Black Soldier Fly (*Hermetia illucens*) larvae have been shown to be able to rapidly consume large amounts of organic waste when kept at 31.8 °C, the optimum temperature for reproduction. Enthusiasts have experimented with a large number of different waste products and some even sell starter kits to the public.

Cockroach Composting

Cockroach composting is another insect-mediated composting method. In this case the adults of any number of cockroach species (such as the Turkestan cockroach or *Blaptica dubia*) are used to quickly convert manure or kitchen waste to nutrient dense compost. Depending on species used and environmental conditions, excess composting insects can be used as an excellent animal feed for farm animals and pets.

Bokashi

Inside a recently started bokashi bin. The aerated base is just visible through the food scraps and bokashi bran.

Bokashi is a method that uses a mix of microorganisms to cover food waste or wilted plants to decrease smell. Bokashi (ぼかし) is Japanese for "shading off" or "gradation." It derives from the practice of Japanese farmers centuries ago of covering food waste with rich, local soil that contained the microorganisms that would ferment the waste. After a few weeks, they would bury the waste.

Most practitioners obtain the microorganisms from the product Effective Microorganisms (EM1), first sold in the 1980s. EM1 is mixed with a carbon base (e.g. sawdust or bran) that it sticks to and a sugar for food (e.g. molasses). The mixture is layered with waste in a sealed container and after a few weeks, removed and buried.

Newspaper fermented in a lactobacillus culture can be substituted for bokashi bran for a successful bokashi bucket.

Compost tea

Compost teas are defined as water extracts brewed from composted materials and can be derived from aerobic or anaerobic processes. Compost teas are generally produced from adding one volume of compost to 4-10 volumes of water, but there has also been debate about the benefits of aerating the mixture. Field studies have shown the benefits of adding compost teas to crops due to the adding of organic matter, increased nutrient availability and increased microbial activity. They have also been shown to have an effect on plant pathogens.

Composting Toilets

A composting toilet does not require water or electricity, and when properly managed does not smell. A composting toilet collects human excreta which is then added to a compost heap together with sawdust and straw or other carbon rich materials, where pathogens are destroyed to some extent. The amount of pathogen destruction depends on the temperature (mesophilic or thermophilic conditions) and composting time. A composting toilet tries to process the excreta in situ although this is often coupled with a secondary external composting step. The resulting compost product has been given various names, such as humanure and EcoHumus.

A composting toilet can aid in the conservation of fresh water by avoiding the usage of potable water required by the typical flush toilet. It further prevents the pollution of ground water by controlling the fecal matter decomposition before entering the system. When properly managed, there should be no ground contamination from leachate.

COMPOST AND LAND-FILLING

As concern about landfill space increases, worldwide interest in recycling by means of composting is growing, since composting is a process for converting decomposable organic materials into useful stable products. Composting is one of the only ways to revitalize soil vitality due to phosphorus depletion in soil.Industrial scale composting in the form of in-vessel composting, aerated static pile composting, and anaerobic digestion takes place in most Western countries now, and in many areas is mandated by law. There are process and product guidelines in Europe that date to the early 1980s (Germany, the Netherlands, Switzerland) and only more recently in the UK and the US. In both these countries, private trade associations within the industry have established loose standards, some say as a stop-gap measure to discourage independent government agencies from establishing tougher consumer-friendly standards. The USA is the only Western country that does not distinguish sludge-source compost from green-composts, and by default in the USA 50% of states expect composts to comply in some manner with the federal EPA 503 rule promulgated in 1984 for sludge products. Compost is regulated in Canada and Australia as well.

Industrial Systems

A large compost pile that is steaming with the heat generated by thermophilic microorganisms.

Industrial composting systems are increasingly being installed as a waste management alternative to landfills, along with other advanced waste processing systems. Mechanical sorting of mixed waste streams combined with anaerobic digestion or in-vessel composting is called mechanical biological treatment, and is increasingly being used in developed countries due to regulations controlling the amount of organic matter allowed in landfills. Treating biodegradable waste before it enters a landfill reduces global warming from fugitive methane; untreated waste breaks down anaerobically in a landfill, producing landfill gas that contains methane, a potent greenhouse gas.

Vermicomposting, also known as vermiculture, is used for medium-scale on-site institutional composting, such as for food waste from universities and shopping malls. It is selected either as a more environmentally friendly choice than conventional methods of disposal, or to reduce the cost of commercial waste removal.

Large-scale composting systems are used by many urban areas around the world. Co-composting is a technique that combines solid waste with de-watered biosolids, although difficulties controlling inert and plastics contamination from municipal solid waste makes this approach less attractive. The world's largest MSW co-composter is the Edmonton Composting Facility in Edmonton, Alberta, Canada, which turns 220,000 tonnes of residential solid waste and 22,500 dry tonnes of biosolids per year into 80,000 tonnes of compost. The facility is 38,690 m^2 (416,500 sq.ft.) in area, equivalent to 4½ Canadian football fields, and the operating structure is the largest stainless steel building in North America, the size of 14 NHL rinks. In 2006, Qatar awarded Keppel Seghers Singapore, a subsidiary of Keppel Corporation, a contract to begin construction on a

275,000 tonne/year anaerobic digestion and composting plant licensed by Kompogas (de) Switzerland. This plant, with 15 independent anaerobic digesters, will be the world's largest composting facility once fully operational in early 2011 and forms part of Qatar's Domestic Solid Waste Management Centre, the largest integrated waste management complex in the Middle East.

Another large MSW composter is the Lahore Composting Facility in Lahore, Pakistan, which has a capacity to convert 1,000 tonnes of municipal solid waste per day into compost. It also has a capacity to convert substantial portion of the intake into refuse-derived fuel (RDF) materials for further combustion use in several energy consuming industries across Pakistan, for example in cement manufacturing companies where it is used to heat cement kilns. This project has also been approved by the Executive Board of the United Nations Framework Convention on Climate Change for reducing methane emissions, and has been registered with a capacity of reducing 108,686 tonnes carbon dioxide equivalent per annum.

RELATED TECHNOLOGIES

Anaerobic digestion is process for converting organic waste into (biogas). The residual material, sometimes in combination with sewage sludge can be followed by an aerobic composting process before selling or giving away the compost.

HISTORY

Compost Basket.

Composting as a recognized practice dates to at least the early Roman Empire since Pliny the Elder (AD 23-79). Traditionally, composting involved piling organic materials until the next planting season, at which time the materials would have decayed enough to be ready for use in the soil. The advantage of this method is that little working time or effort is required from the composter and it fits in naturally with agricultural practices in temperate climates. Disadvantages (from the modern perspective) are that space is used for a whole year, some nutrients might be leached due to exposure to rainfall, and disease-producing organisms and insects may not be adequately controlled.

Composting was somewhat modernized beginning in the 1920s in Europe as a tool for organic farming. The first industrial station for the transformation of urban organic materials into compost was set up in Wels, Austria in the year 1921. Early frequent citations for propounding composting within farming are for the German-speaking world Rudolf Steiner, founder of a farming method called biodynamics, and Annie Francé-Harrar, who was appointed on behalf of the government in Mexico and supported the country 1950–1958 to set up a large humus organization in the fight against erosion and soil degradation.

In the English-speaking world it was Sir Albert Howard who worked extensively in India on sustainable practices and Lady Eve Balfour who was a huge proponent of composting. Composting was imported to America by various followers of these early European movements by the likes of J.I. Rodale (founder of Rodale Organic Gardening), E.E. Pfeiffer (who developed scientific practices in biodynamic farming), Paul Keene (founder of Walnut Acres in Pennsylvania), and Scott and Helen Nearing (who inspired the back-to-the-land movement of the 1960s). Coincidentally, some of the above met briefly in India - all were quite influential in the U.S. from the 1960s into the 1980s.

There are many modern proponents of rapid composting that attempt to correct some of the perceived problems associated with traditional, slow composting. Many advocate that compost can be made in 2 to 3 weeks. Many such short processes involve a few changes to traditional methods, including smaller, more homogenized pieces in the compost, controlling carbon-to-nitrogen ratio (C:N) at 30 to 1 or less, and monitoring the moisture level more carefully. However, none of these parameters differ significantly from the early writings of Howard and Balfour, suggesting that in fact modern composting has not made significant advances over the traditional methods that take a few months to work. For this reason and others, many modern scientists who deal with carbon transformations are sceptical that there is a “super-charged” way to get nature to make compost rapidly.

In fact, both sides are right to some extent. The bacterial activity in rapid high heat methods breaks down the material to the extent that pathogens and seeds are destroyed, and the original feedstock is unrecognizable. At this stage, the compost can be used to prepare fields or other planting areas. However, most professionals recommend that the compost be given time to cure before using in a nursery for starting seeds or growing young plants. The curing time allows fungi to continue the decomposition process and eliminating phytotoxic substances.

Many countries such as Wales and some individual cities such as Seattle and San Francisco require food and yard waste to be sorted for composting.

Kew Gardens in London has one of the biggest non-commercial compost heaps in Europe.

COMPOSTING TOILET

Composting toilet at Activism Festival 2010 in the mountains outside Jerusalem.

A **composting toilet** is a type of dry toilet that uses a predominantly aerobic processing system to treat human excreta, by composting or managed aerobic decomposition. These toilets generally use little to no water and may be used as an alternative to flush toilets. They have found use in situations where no suitable water supply or sewer system and sewage treatment plant is available to capture the nutrients in human excreta. They are in use in many roadside facilities and national parks in Sweden, Canada, US, UK and Australia. They are used in rural holiday homes in Sweden and Finland.

The human excreta is usually mixed with sawdust, coconut coir or peat moss to facilitate aerobic processing, liquid absorption, and odor mitigation. Most composting toilets use slow, cold composting conditions, sometimes connected to a secondary external composting step.

Composting toilets produce a compost that may be used for horticultural or agricultural soil enrichment if the local regulations allow this. A curing stage is often needed to allow mesophilic composting to reduce potential phytotoxins.

Terminology

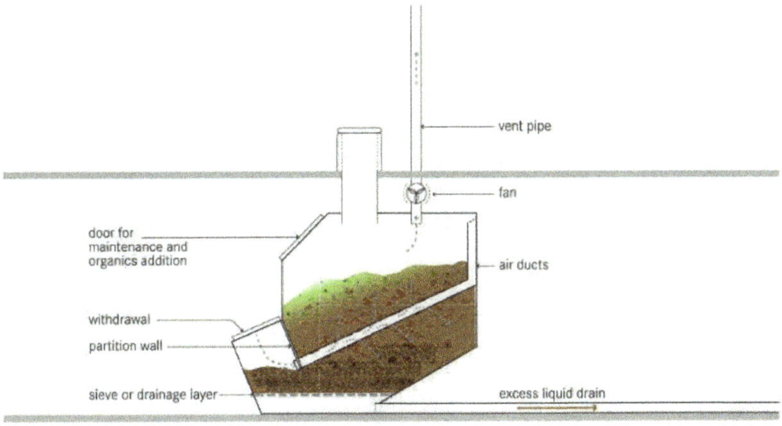

Schematic of the composting chamber which is located below the toilet seat

The term "composting toilet" is used quite loosely, and its meaning may vary by country. For example, in English-speaking countries, the term "anaerobic composting" (equivalent to anaerobic decomposition) is used. In Germany and Scandinavian countries, composting always refers to a predominantly aerobic process. This aerobic composting may take place with an increase in temperature due to microbial action, or without a temperature increase in the case of slow composting or cold composting. If earth worms are used (vermicomposting) then there is also no increase in temperature.

Composting toilets differ from pit latrines, arborloo or tree bogs, all of which are forms of less controlled decomposition and may not protect groundwater from nutrient or pathogen contamination or provide optimal nutrient recycling. They also differ from urine-diverting dry toilets (UDDTs) where pathogen reduction is achieved through dehydration (also known by the more precise term "desiccation") and where the faeces collection vault is kept as dry as possible. Composting toilets target a certain degree of moisture in the composting chamber.

Composting toilets usually do not divert urine. Offering a waterless urinal in addition to the toilet can help keep excess amounts of urine out of the composting chamber.

Composting toilets can be used to implement an ecological sanitation approach for resource recovery, and some people call their composting toilet designs «ecosan toilets» for that reason. However, this is not recommended as the two terms (i.e. composting and ecosan) are not identical.

Composting toilets have also been called «sawdust toilets», which can be appropriate if the amount of aerobic composting taking place in the toilet›s container is very limited. The «Clivus multrum» is a type of composting toilet which has a large composting chamber below the toilet seat and also receives undigested organic material to increase the carbon to nitrogen ratio.

Applications

This is the pedestal for a split-system composting toilet where collection/treatment chambers are located below the bathroom floor.

Inexpensive do-it-yourself compost toilet at Dial House, Essex, England, utilizing an old desk as the toilet unit.

Public composting toilet at a highway rest facility in Sweden.

Composting toilets can be suitable in areas such as a rural area or a park that lacks a suitable water supply, sewers and sewage treatment. They can also help increase the resilience of existing sanitation systems in the face of possible natural disasters such as climate change, earthquakes or tsunami. Composting toilets can reduce or perhaps eliminate the need for a septic tank system to reduce environmental footprint (particularly when used in conjunction with an on-site greywater treatment system).

These types of toilets can be used for resource recovery by reusing sanitized feces and urine as fertilizer and soil conditioner for gardening or ornamental activities.

Basics

Components

A composting toilet consists of two elements: a place to sit or squat and a collection/ composting unit. The composting unit consists of four main parts:

- storage or composting chamber
- a ventilation unit to ensure that the degradation process in the toilet is predominantly aerobic and to vent odorous gases
- a leachate collection system to remove excess liquid
- an access door for extracting the compost

Construction

The composting chamber can be constructed above or below ground level. It can be inside a structure or include a separate superstructure.

A drainage system removes leachate. Otherwise, excess moisture can cause anaerobic conditions and impede decomposition. Urine diversion can improve compost quality, since urine contains large amounts of ammonia that inhibits microbiological activity.

Composting toilets greatly reduce human waste volumes through psychrophilic, thermophilic or mesophilic composting. Keeping the composting chamber insulated and warm protects the composting process from slowing due to low temperatures.

Odorous Gases

The following gases may be emitted during the composting process that takes place in composting toilets: hydrogen sulfide (H_2S), ammonia, nitrous oxide (N_2O) and volatile organic compounds (VOCs). These gases can potentially lead to complaints about odours. Some methane may also be present, but it is not odorous.

Pathogen Removal

Excreta-derived compost recycles fecal nutrients, but it can carry and spread pathogens if the process of reuse of excreta is not done properly.

Internal pathogen destruction rates are usually low, particularly helminth eggs, such as Ascaris eggs. This carries the risk of spreading disease if a proper system management is not in place. Compost from human excreta processed under only mesophilic conditions or taken directly from the compost chamber is not safe for food production. High temperatures or long composting times are required to kill helminth eggs, the hardiest of all pathogens. Helminth infections are common in many developing countries.

In thermophilic composting bacteria that thrive at temperatures of 40–60 °C (104–140 °F) oxidize (break down) waste into its components, some of which are consumed in the process, reducing volume and eliminating potential pathogens. To destroy pathogens, thermophilic composting must heat the compost pile sufficiently, or enough time (1–2 years) must elapse since fresh material was added that biological activity has had the same pathogen removal effect.

One guideline claims that pathogen levels are reduced to a safe level by thermophilic composting at temperatures of 55 °C for at least two weeks or at 60 °C for one week. An alternative guideline claims that complete pathogen destruction may be achieved already if the entire compost heap reaches a temperature of 62 °C (144 °F) for one hour, 50 °C (122 °F) for one day, 46 °C (115 °F) for one week or 43 °C (109 °F) for one month, although others regard this as overly optimistic.

Design Considerations

Composting toilet with a seal in the lid in Germany.

Environmental Factors

Four main factors affect the decomposition process:

- Sufficient oxygen is necessary for aerobic composting
- Moisture content from 45 to 70 percent (heuristically, "the compost should feel damp to the touch, with only a drop or two of water expelled when tightly squeezed in the hand.")
- Temperature between 40 and 50 °C (achieved through proper chamber dimensioning and possibly active mixing)
- Carbon-to-nitrogen ratio (C:N) of 25:1

Additives and Bulking Material

Human excreta and food waste do not provide optimum conditions for composting. Usually the water and nitrogen content is too high, particularly when urine is mixed with feces. Additives or "bulking material", such as wood chips, bark chips, sawdust, ash and pieces of paper can absorb moisture. The additives improve pile aeration and increase the carbon to nitrogen ratio. Bulking material also covers faeces and reduces insect access. Absent sufficient bulking material, the material may become too compact and form impermeable layers, which leads to anaerobic conditions and odour.

Leachate management

Leachate removal controls moisture levels, which is necessary to ensure rapid, aerobic composting. Some commercial units include a urine-separator or urine-diverting system and/or a drain at the bottom of the composter for this purpose.

Aeration and mixing

Microbial action also requires oxygen, typically from the air. Commercial systems provide ventilation that moves air from the bathroom, through the waste container, and out a vertical pipe, venting above the roof. This air movement (via convection or fan forced) passes carbon dioxide and odors.

Some units require manual methods for periodic aeration of the solid mass such as rotating the composting chamber or pulling an "aerator rake" through the mass.

Types

External composting chamber of a composting toilet at a house in France.

Commercial units and construct-it-yourself systems are available. Variations include number of composting vaults, removable vault, urine diversion and active mixing/aeration.

Slow Composting (or moldering) Toilets

Most composting toilets use slow composting which is also called "cold composting". The compost heap is built up step by step over time.

The finished end product from "slow" composting toilets ("moldering toilets" or "moldering privies" in the US), is generally not free of pathogens. World Health Organization Guidelines from 2006 offer a framework for safe reuse of excreta, using a multiple barrier approach.

Slow composting toilets employ a passive approach. Common applications involve modest and often seasonal use, such as remote trail networks. They are typically designed such that the materials deposited can be isolated from the operational part. The toilet can also be closed to allow further mesophilic composting. Slow composting toilets rely on long retention times for pathogen reduction and for decomposition of excreta or on the combination of time and/or the addition of red wriggler worms for vermi-composting. Worms can be introduced to accelerate composting. Some jurisdictions of the US consider these worms as invasive species.

Example in Vermont Woods

Slow composting toilets have been installed by the Green Mountain Club in Vermont's woodlands. They employ multiple vaults (called cribs) and a movable building. When one of the vaults fills, the building is moved over an empty vault. The full vault is left untouched for as long as possible (up to three years) before it is emptied. The large surface area and exposure to air currents can cause the pile to dry out. To counteract this, signs instruct users to urinate in the toilet. The club also uses pit latrines and simple bucket toilets with woodchips and external composting and directs users to urinate in the forest to prevent odiferous anaerobic conditions.

Active Composters

Self-Contained

"Self-contained" composting toilets compost in a container within the toilet unit. They are slightly larger than a flush toilet, but use roughly the same floor space. Some units use fans for aeration, and optionally, heating elements to maintain optimum temperatures to hasten the composting process and to evaporate urine and other moisture. Operators of composting toilets commonly add a small amount of absorbent carbon material (such as untreated sawdust, coconut coir, peat moss) after each use to create air pockets to encourage aerobic processing, to absorb liquid and to create an odor barrier. This additive is sometimes referred to as "bulking agent." Some owner-operators use microbial "starter" cultures to ensure composting bacteria are in the process, although this is not critical.

Remote

"Remote" "central" or "underfloor" units collect excreta via a toilet stool, either waterless, vacuum or micro-flush, from which it drains into a composter. "Vacuum-flush systems" can flush horizontally or upward with a small amount of water to the composter. "Micro-flush" toilets use about 500 millilitres (17 US fl oz) per use. These units feature a chamber below the toilet stool (such as in a basement or outside) where composting takes place and are suitable for high-volume and year-round applications as well as to serve multiple toilet stools.

Other

Some units employ roll-away containers fitted with aerators, while others use sloped-bottom tanks.

Maintenance

Maintenance is critical to ensure proper operation, including odor prevention. Maintenance tasks include: cleaning, servicing technical components such as fans and removal of compost, leachate and urine. Urine removal is only required for those types of composting toilets using urine diversion.

Once composting is complete (or more often), the compost must be removed from the unit. How often this occurs is a function of container size, usage and composting conditions, such as temperature. Active, hot composting may span months only while passive, cold composting may require years. Properly managed units yield output volumes of about 10% of inputs.

Uses of Compost

The material from composting toilets is a humus-like material, which can be suitable as a soil amendment for agriculture. Compost from residential composting toilets can be used in domestic gardens, and this is the main such use.

Enriching soil with compost adds substantial nitrogen, phosphorus, potassium, carbon and calcium. In this regard compost is equivalent to many fertilizers and manures purchased in garden stores. Compost from composting toilets has a higher nutrient availability than the dried faeces that result from a urine-diverting dry toilet.

Finished compost from a composting toilet ready for application as soil improvement in Kiel-Hassee, Germany

Urine is typically present, although some is lost via leaching and evaporation. Urine can contain up to 90 percent of the residual nitrogen, up to 50 percent of the phosphorus, and up to 70 percent of the potassium.

Compost derived from these toilets has in principle the same uses as compost derived from other organic waste products, such as sewage sludge or municipal organic waste. However, users of excreta-derived compost must consider the risk of pathogens.

Pharmaceutical Residues

Excreta-derived compost may contain prescription pharmaceuticals. Such residues are also present in conventional wastewater treatment effluent. This could contaminate groundwater. Among the medications that have been found in groundwater in recent years are antibiotics, antidepressants, blood thinners, ACE inhibitors, calcium-channel blockers, digoxin, estrogen, progesterone, testosterone, Ibuprofen, caffeine, carbamazepine, fibrates and cholesterol-reducing medications. Between 30% and 95% of pharmaceuticals medications are excreted by the human body. Medications that are lipophilic (dissolved in fats) are more likely to reach groundwater by leaching from fecal wastes. Wastewater treatment plants remove an average of 60% of these medications. The percentage of medications degraded during composting of excreta has not yet been reported.

Comparison

Pit Latrines

Unlike pit latrines, composting toilets convert feces into a dry, odorless material, avoiding the issues surrounding liquid fecal sludge management (e.g. odor, insects and disposal). These toilets minimize the risk of water pollution through the safe containment of feces in above-ground vaults, which allows the toilets to be sited in locations where pit-based systems are not appropriate.

However, composting toilets face higher capital costs (although lifecycle costs might be lower) and greater complexity (for instance, adding covering materials and managing moisture content).

Flush Toilets

Unlike flush toilets, composting toilets do not dilute excreta and create wastewater streams which must be treated before disposal. On the other hand, wastewater treatment plants can centralize waste management for an entire community, with potentially greater efficiency.

Urine-Diverting Dry Toilets

Composting toilets are more difficult to maintain than other types of dry toilets, like urine-diverting dry toilets (UDDT) with which they are often confused. This is due to the need to maintain a consistent and relatively high moisture content, as well as the relatively high complexity of composting toilets compared to UDDTs. Apart from that, composting toilets are quite similar to UDDTs, sharing many of the same advantages and disadvantages.

Henry Moule's earth closet, patented in 1873 (not a true composting toilet). Example from around 1875. Rear chamber for dispensing cover material.

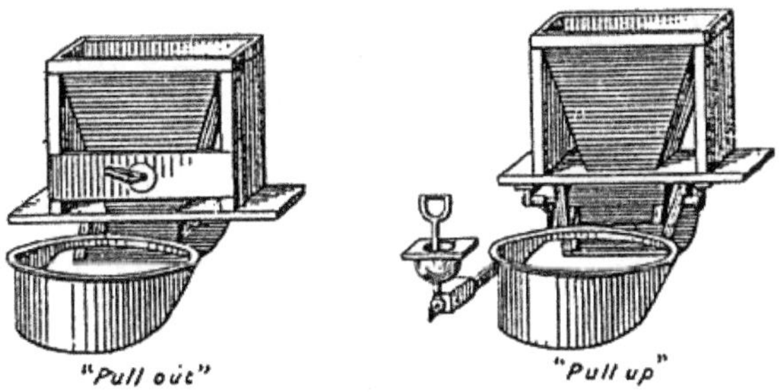

FIGS. 8 and 9.—Moule's Earth Closets.

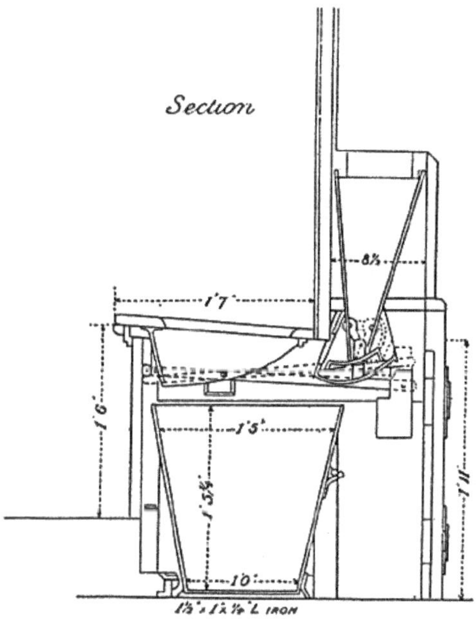

FIG. 10.—Section of Moule's Automatic Earth Closet.

Henry Moule's earth closet design, circa 1909.

Dry Earth Toilet

Before the flush toilet became accepted in the late 19th century in developed countries, some inventors, scientists and public health officials supported the use of "dry earth closets", a type of dry toilet with similarities to composting toilets, but the collection vessel for the human excreta was not designed to compost. Dry earth closets were invented by English clergyman Henry Moule, who dedicated his life to improving public sanitation after witnessing the cholera epidemics of 1849 and 1854. Impressed by the insalubrity of the houses, especially during the Great Stink in the summer of 1858, he invented what he called the 'dry earth system'.

In partnership with James Bannehr, he patented his device (No. 1316, dated 28 May 1860). Among his works bearing on the subject were *The Advantages of the Dry Earth System* (1868), *The Impossibility overcome: or the Inoffensive, Safe, and Economical*

*Disposal of the Refuse of Towns and Villages (*1870), The *Dry Earth System* (1871), *Town Refuse, the Remedy for Local Taxation* (1872), and *National Health and Wealth promoted by the general adoption of the Dry Earth System* (1873).

His system was adopted in private houses, in rural districts, in military camps, in many hospitals, and extensively in the British Raj. Ultimately, however, it failed to gain public support as attention turned to the water-flushed toilet connected to a sewer system.

In Germany, a similar dry toilet with a peat dispenser was marketed until after the second World War (it was called «Metroclo» and was manufactured by Gefinal, Berlin).

Society and Culture

Regulations

International Organization for Standardization (ISO)

The International Organization for Standardization (ISO) is currently preparing a "management standard". As of 2015 this was in a draft state as ISO 24521, under the heading "Activities relating to drinking water and wastewater services — Guidelines for the management of basic onsite domestic wastewater services". The standard is meant to be used in conjunction with ISO 24511. It deals with toilets (including composting toilets) and toilet waste. The guidelines are applicable to basic wastewater systems and include the complete domestic wastewater cycle, such as planning, usability, operation and maintenance, disposal, reuse and health.

International Association of Plumbing and Mechanical Officials

The International Association of Plumbing and Mechanical Officials (IAPMO) is a plumbing and mechanical code structure adopted by many developed countries. It recently proposed an addition to its "Green Plumbing Mechanical Code Supplement" that, "... outlines performance criteria for site built composting toilets with and without urine diversion and manufactured composting toilets." If adopted, this composting and urine diversion toilet code (the first of its kind in the United States) will appear in the 2015 edition of the Green Supplement to the Uniform Plumbing Code.

United States

No performance standards for composting toilets are universally accepted in the US. Seven jurisdictions in North Americause *American National Standard/NSF International Standard ANSI/NSF 41-1998: Non-Liquid Saturated Treatment Systems*. An updated version was published in 2011. Systems might also be listed with the Canadian Standards Association, cETL-US, and other standards programs.

Regarding byproduct regulation, several US states permit disposal of solids from composting toilets (usually a distinction between different types of dry toilets is not made) by burial, with varying or no minimum depth mandates (as little as 6 inches). For instance:

Massachusetts: "Residuals from the composting toilet system must be buried on-site and covered with a minimum of six inches of clean compacted soil.Massachusetts requires that any liquids produced but, "not recycled through the toilet itself be either discharged through a greywater system on the property that includes a septic tank and soil absorption system, or removed by a licensed septage hauler."

- Oregon: "Humus from composting toilets may be used around ornamental shrubs, flowers, trees, or fruit trees and shall be buried under at least twelve inches of soil cover."
- Rhode Island: "Solids produced by alternative toilets may be buried on site," while, "residuals shall not be applied to food crops."
- Virginia: "All materials removed from a composting privy shall be buried," and "compost material shall not be placed in vegetable gardens or on the ground surface."
- Vermont: "Byproducts may be disposed via "...shallow burial in a location approved by the Agency that meets the minimum site conditions required for an onsite septic tank-based sanitation system."
- Washington: models its extensive regulations for what it refers to as "waterless toilets" on the federal regulations that govern sewage sludge.

The Environmental Protection Agency has no jurisdiction over the byproducts of a dry toilet as long as excreta are not referred to as "fertilizer" (but instead simply a material that is being disposed of). Federal rule 503, known colloquially as the "EPA Biosolids rule" or the "EPA sludge rule" applies only to fertilizer. Thus, individual states regulate composting toilets.

Germany

The regulations for composting toilets and other forms of dry toilets in Germany vary from state to state and from one application to another (e.g. use in allotment gardens or use in family homes and settlements). In the different states of Germany, it is the "Landesbauordnung" (translates to "state civil engineering regulations") of the respective state that regulates the use of such alternative toilets. Most of them stipulate the use of flush toilets, however there are many exceptions, for example in the states of Hamburg, Lower Saxony, Bavaria, Mecklenburg-Western Pomerania, Rhineland-Palatinate, Saxony-Anhalt and Thuringia. These generally make exceptions for the use of composting toilets in homes provided that there are no concerns for public health.

Regulations governing the use of compost and urine from composting toilets is less clear in Germany but it seems generally allowed provided it is used on one's own property and not sold to third parties.

Examples

Finland

Numerous sparsely settled villages in rural areas in Finland are not connected to municipal water supply or sewer networks, requiring homeowners to operate their own systems. Individual private wells, i.e. shallow dug wells or boreholes in the bedrock, are often used for water supply, and many homeowners have opted for composting toilets. In addition, these toilets are common at holiday homes, often located near sensitive water bodies. For these reasons, many manufacturers of composting toilets are based in Finland, including Biolan, Ekolet, Kekkilä, Pikkuvihreä and Raita Environment.

Estimates made by leading Finnish composting toilet manufacturers and the Global Dry Toilet Association of Finland provided the following 2014 figures for composting toilet use in Finland:

- About 4% of single-family homes not connected to a public sewer network are equipped with a composting toilet.
- Some 200,000 manufactured composting toilets are thought to serve holiday homes, matched by the number of other dry toilets. The simplest ones are sited in an outhouse.

Germany

Composting container of "TerraNova" composting toilet, showing open removal chamber (town house at the ecological settlement Hamburg-Allermöhe, Germany).

Composting toilets have been successfully installed in houses with up to four floors. An estimate from 2008 put the number of composting toilets in households in Germany at 500. Most of these residences are also connected to a sewer system; the composting toilet was not installed due a lack of sewer system but for other reasons, mainly because of an "ecological mindset" of the owners.

In Germany and Austria, composting toilets and other types of dry toilets have been installed in single and multi-family houses (e.g. Hamburg, Freiburg, Berlin), ecological settlements (e.g. Hamburg-Allemöhe, Hamburg-Braamwisch, Kiel-Hassee, Bielefeld-Waldquelle, Wien-Gänserndorf) and in public buildings (e.g. Ökohaus Rostock, VHS-Ökostation Stuttgart-Wartberg, public toilets in recreational areas, restaurants and huts in the Alps, house boats and forest Kindergartens).

The ecological settlement in Hamburg-Allermöhe has had composting toilets since 1982. The settlement of 36 single-family houses with approximately 140 inhabitants uses composting toilets, rainwater harvesting and constructed wetlands. Composting toilets save about 40 litres of water per capita per day compared to a conventional flush toilet (10 liter per flush), which adds up to 2,044 m^3 water savings per year for the whole settlement.

Worldwide

Composting toilets with a large composting container (of the type Clivus Multrum and derivations of it) are popular in US, Canada, Australia, New Zealand and Sweden. They can be bought and installed as commercial products, as designs for self builders or as «design derivatives» which are marketed under various names. It has been estimated that approximately 10,000 such toilets might be in use worldwide.

REFERENCES

1. Tilley, E.; Ulrich, L.; Lüthi, C.; Reymond, Ph.; Zurbrügg, C. Compendium of Sanitation Systems and Technologies - (2nd Revised Edition). Swiss Federal Institute of Aquatic Science and Technology (Eawag), Duebendorf, Switzerland. p. 72. ISBN 978-3-906484-57-0.
2. Berger, W. (2011). Technology review of composting toilets - Basic overview of composting toilets (with or without urine diversion). Deutsche Gesellschaft für Internationale Zusammenarbeit (GIZ) GmbH, Eschborn, Germany
3. Rieck, C., von Münch, E., Hoffmann, H. (2012). Technology review of urine-diverting dry toilets (UDDTs) - Overview on design, management, maintenance and costs. Deutsche Gesellschaft fuer Internationale Zusammenarbeit (GIZ) GmbH, Eschborn, Germany
4. Hill, B. G. (2013). An evaluation of waterless human waste management systems at North American public remote sites. PhD thesis, University of British Columbia (Vancouver), Canada
5. "The online Compendium of Sanitation Systems and Technologies". The online Compendium of Sanitation Systems and Technologies. eawag aquatic research. 2014. Retrieved 2014-12-29.

6. Font, Xavier; Artola, Adriana; Sánchez, Antoni (6 April 2011). "Detection, Composition and Treatment of Volatile Organic Compounds from Waste Treatment Plants". Sensors. 11 (12): 4043–4059. doi:10.3390/s110404043.
7. Stenström, T.A., Seidu, R., Ekane, N., Zurbrügg, C. (2011). Microbial exposure and health assessments in sanitation technologies and systems - EcoSanRes Series, 2011-1. Stockholm Environment Institute (SEI), Stockholm, Sweden, page 88
8. National Small Flows Clearinghouse, West Virginia University, Composting toilet technology
9. WHO (2006). WHO Guidelines for the Safe Use of Wastewater, Excreta and Greywater - Volume IV: Excreta and greywater use in agriculture. World Health Organization (WHO), Geneva, Switzerland
10. Appalachian Trail Conservancy (2014). Backcountry Sanitation Manual, 2nd Edition. Appalachian Trail Conservancy, Green Mountain Club, USDA Forest Service, National Park Service, USA
11. Allen, Lee (2013). "Long Trail News: Quarterly of the Green Mountain Club, Fall 2013. Article titled: "A Privy is a Privy is a Privy...or is it? To Pee or Not Pee.""(PDF). Green Mountain Club. Green Mountain Club. Retrieved 31 January 2013.
12. Antos-Ketcham, Pete (2013). "Long Trail News: Quarterly of the Green Mountain Club, Fall 2013. Article titled: "Batch-Bin/Beyond-the-Bin (BTB) Composting Privies"" (PDF). Green Mountain Club. Green Mountain Club. Retrieved 31 January 2015.
13. Berger, W. (2009). Appendix of technology review of composting toilets - List of manufacturers and commercially available composting toilets. Gesellschaft für Internationale Zusammenarbeit (GIZ) GmbH
14. J.O. Drangert, Urine separation systems
15. Drugs in the Water. Harvard Health Letter. 2011.
16. Encyclopedia of Quantitative Risk Analysis and Assessment, Volume 1, edited by Edward L. Melnick, Brian S. Veritt, 2008
17. "ISO/DIS 24521. Activities relating to drinking water and wastewater services -- Guidelines for the management of basic onsite domestic wastewater services". International Organization for Standardization (ISO). Retrieved 15 January 2015.
18. "ISO 24511:2007. Activities relating to drinking water and wastewater services -- Guidelines for the management of wastewater utilities and for the assessment of wastewater services". International Organization for Standardization (ISO). Retrieved 15 January 2015.
19. "Recode September 2014 Newsletter". Recode. Recode. September 2014. Retrieved 15 January 2015.
20. "IAPMO Proposed Composting and Urine DIversion Toilet Code" (PDF). The IAPMO Group. International Association of Plumbing and Mechanical Officials. Retrieved 15 January 2015.

21. Cole, Daniel (January 2015). "IAPMO GPMCS raising the bar for water, energy efficiency". Plumbing Engineer. Plumbing Engineer. Retrieved 15 January 2015.
22. Oregon Onsite Advisory Committee "Final Report of Recommended Changes to Rules Governing Onsite Systems", OR DEQ, February 8, 2010, accessed May 8, 2011.
23. "PUBLICATIONS - Standards and Criteria - March 21, 2013" (PDF). NSF International. p. 4. Retrieved 24 March 2013. Wastewater Treatment Units ... NSF/ANSI 41 – 2011: Non-liquid saturated treatment systems (composting toilets)
24. "Regulatory Provisions for Composting Toilets and Greywater Systems". The Official Website of the Massachusetts Executive Office of Energy and Environmental Affairs. Office of Energy and Environmental Affairs. Retrieved 13 January 2015.
25. "Department of Consumer and Business Services, Building Codes Division, Division 770, Plumbing Product Approvals". Oregon Secretary of State. State of Oregon. Retrieved 13 January 2015.
26. "State of Rhode Island and Providence Plantations Department of Environmental Management, Office of Water Resources: "Rules Establishing Minimum Standards Relating to Location, Design, Construction and Maintenance of Onsite Wastewater Treatment Systems"" (PDF). State of Rhode Island Department of Environmental Management. STATE OF RHODE ISLAND AND PROVIDENCE PLANTATIONS. July 2010. Retrieved 13 January 2015.
27. "SEWAGE HANDLING AND DISPOSAL REGULATIONS (Emergency Regulations for Gravelless Material and Drip Dispersal), 12 VAC 5-610-10 et seq." (PDF). State of Virginia Department of Health. Commonwealth of Virginia. 14 March 2014. Retrieved 13 January 2015.
28. "Environmental Protection Rules, Chapter 1: Wastewater System and Potable Water Supply Rules" (PDF). State of Vermont Drinking Water and Groundwater Protection Division. State of Vermont. 29 September 2007. Retrieved 14 January 2015.
29. "Recommended Standards and Guidance for Performance, Application, Design, and Operation & Maintenance: Water Conserving On-Site Wastewater Treatment Systems" (PDF). State of Washington Department of Health. State of Washington. July 2012. Retrieved 14 January 2015.
30. "Water Efficiency Technology Fact Sheet: Composting Toilets" (PDF). United States Environmental Protection Agency, Office of Water, Washington, D.C., EPA 832-F-99-066. United States Environmental Protection Agency, Office of Water. September 1999. Retrieved 13 January 2015.
31. "TITLE 40—Protection of Environment, Chapter I—Environmental Protection Agency (Continued), Subchapter O—Sewage Sludge, Part 503—Standards for the Use or Disposal of Sewage Sludge". Electronic Code of Federal Regulations. United States Government Publishing Office. Retrieved 13 January 2015.

32. Lorenz-Ladener, Hrsg. Claudia; Berger, Wolfgang (2005). Kompost-Toiletten: Wege zur sinnvollen Fäkalienentsorgung (1. überarb. u. erw. Aufl. ed.). Staufen im Breisgau: Ökobuch. p. 178. ISBN 978-3-936896-16-9.
33. Global Dry Toilet Association of Finland (2011) Dry Toilet Manufacturers in Finland, Leaflet in English and Finnish
34. «Global Dry Toilet Association of Finland». Global Dry Toilet Association of Finland - Company and association members. Retrieved 15 January 2015.
35. Lorenz-Ladener, Hrsg. Claudia; Berger, Wolfgang (2005). Kompost-Toiletten: Wege zur sinnvollen Fäkalienentsorgung (1. überarb. u. erw. Aufl. ed.). Staufen im Breisgau: Ökobuch. p. 183. ISBN 978-3-936896-16-9.
36. «Are LED Grow Lights Really That Efficient?». Growgear. September 21, 2013. Retrieved November 16, 2016.
37. http://hortsci.ashspublications.org/content/30/2/374.full.pdf
38. Learning centre
39. Plant Growth Factors: Light
40. http://www.mv.helsinki.fi/aphalo/photobio/pdf/notes1.pdf
41. Phototropins Promote Plant Growth in Response to Blue Light in Low Light Environments
42. http://hortsci.ashspublications.org/content/45/2/236.full.pdf
43. Green-light Supplementation for Enhanced Lettuce Growth under Red- and Blue-light-emitting Diodes
44. Green Light Drives Leaf Photosynthesis More Efficiently than Red Light in Strong White Light: Revisiting the Enigmatic Question of Why Leaves are Green
45. Howard Resh (16 January 2013). Hobby Hydroponics, Second Edition. CRC Press. pp. 23–28. ISBN 978-1-4665-6942-3.
46. George F. Van Patten (2002). Gardening Indoors: The Indoor Gardener›s Bible. Van Patten Publishing. pp. 47–50. ISBN 978-1-878823-31-1.
47. Max Clarke (13 March 2013). The Complete Guide to Building Your Own Greenhouse. Lulu.com. p. 53. ISBN 978-1-105-91367-9.
48. Ellen Zachos (16 December 2014). Growing Healthy Houseplants: Choose the Right Plant, Water Wisely, and Control Pests. Storey Publishing, LLC. pp. 12–13. ISBN 978-1-61212-440-7.
49. Sue Reed (18 October 2013). Energy-Wise Landscape Design: A New Approach for Your Home and Garden. New Society Publishers. pp. 247–250. ISBN 978-1-55092-443-5.
50. Wayne C. Turner; Steve Doty (2007). Energy Management Handbook. The Fairmont Press, Inc. pp. 376–378. ISBN 978-0-88173-543-7.
51. Turner, Wayne C.; Doty, Steve (2007-01-01). Energy Management Handbook. The Fairmont Press, Inc. p. 376. ISBN 9780881735437.
52. http://www.cycloptics.com/sites/default/files/initial_testing_results.pdf

53. Anil Ahuja (9 March 2013). Integrated M/E Design: Building Systems Engineering. Springer Science & Business Media. pp. 28–31. ISBN 978-1-4757-5514-5.

54. «Conversion (MH<-->HPS)». Growers House. Growers House. Retrieved 2015-12-15.

55. «Frequently Asked Questions». leds.hrt.msu.edu. Retrieved 2015-09-19.

56. Yorio, Neil C.; Goins, Gregory D.; Kagie, Hollie R.; Wheeler, Raymond M.; Sager, John C. (2001-04-01). «Improving Spinach, Radish, and Lettuce Growth under Red Light-emitting Diodes (LEDs) with Blue Light Supplementation». HortScience. 36 (2): 380–383. ISSN 0018-5345.

57. Goins, G. D.; Yorio, N. C.; Sanwo, M. M.; Brown, C. S. (1997-07-01). «Photomorphogenesis, photosynthesis, and seed yield of wheat plants grown under red light-emitting diodes (LEDs) with and without supplemental blue lighting». Journal of Experimental Botany. 48 (312): 1407–1413. ISSN 0022-0957. PMID 11541074.

58. Goins, G. D.; Yorio, N. C.; Sanwo-Lewandowski, M. M.; Brown, C. S. (1998-01-01). «Life cycle experiments with Arabidopsis grown under red light-emitting diodes (LEDs)». Life Support & Biosphere Science: International Journal of Earth Space. 5 (2): 143–149. ISSN 1069-9422. PMID 11541670.

59. Massa, Gioia D.; Kim, Hyeon-Hye; Wheeler, Raymond M.; Mitchell, Cary A. (2008-12-01). «Plant Productivity in Response to LED Lighting». HortScience. 43 (7): 1951–1956. ISSN 0018-5345.

60. Kim, Hyeon-Hye; Wheeler, Raymond M.; Sager, John C.; Yorio, Neil C.; Goins, Gregory D. (2005-01-01). «Light-emitting diodes as an illumination source for plants: a review of research at Kennedy Space Center». Habitation (Elmsford, N.Y.). 10 (2): 71–78. doi:10.3727/154296605774791232. ISSN 1542-9660. PMID 15751143.

61. Kim, Hyeon-Hye; Goins, Gregory D.; Wheeler, Raymond M.; Sager, John C. (2004-12-01). «Green-light Supplementation for Enhanced Lettuce Growth under Red- and Blue-light-emitting Diodes». HortScience. 39 (7): 1617–1622. ISSN 0018-5345.

62. Sabzalian, Mohammad R., Parisa Heydarizadeh, Morteza Zahedi, Amin Boroomand, Mehran Agharokh, Mohammad R. Sahba, and Benoît Schoefs. «High performance of vegetables, flowers, and medicinal plants in a red-blue LED incubator for indoor plant production.» Agronomy for Sustainable Development 34, no. 4 (2014): 879-886.

63. Abhay Thosar, Ph.D., Esther Hogeveen van Echtelt (July 29, 2014). LED: the New Fast-track to Growth: Recipe Development and Practical Applications in Horticulture - Part 1: Global Examples LED Recipes and Development (Video). Orlando, FL. Event occurs at 00:03:15, 00:13:05. Archived from the original on February 12, 2015. Retrieved February 12, 2015.

64. Parks, Brian M. (2003-12-01). «The Red Side of Photomorphogenesis». Plant Physiology. 133 (4): 1437–1444. doi:10.1104/pp.103.029702. ISSN 1532-2548. PMC 1540344 . PMID 14681526.

65. «Vertical ‹Pinkhouses:› The Future Of Urban Farming?». NPR.org. Retrieved 2016-02-08.
66. Hoenecke, M. E.; Bula, R. J.; Tibbitts, T. W. (1992-05-01). «Importance of ‹blue› photon levels for lettuce seedlings grown under red-light-emitting diodes». HortScience: A Publication of the American Society for Horticultural Science. 27 (5): 427–430. ISSN 0018-5345. PMID 11537611.
67. Wheeler, R. M.; Mackowiak, C. L.; Sager, J. C. (1991-10-01). «Soybean stem growth under high-pressure sodium with supplemental blue lighting». Agronomy Journal. 83 (5): 903–906. doi:10.2134/agronj1991.00021962008300050024x. ISSN 0002-1962. PMID 11537676.
68. «Evaluation of lettuce growth using supplemental green light with red and blue light-emitting diodes in a controlled environment - a review of research at kennedy space center». www.actahort.org. Retrieved 2016-02-12.
69. Folta, Kevin; Maruhnich, Stefanie A. (2007-09-01). «Green light: a signal to slow down or stop». Journal of Experimental Botany. 58 (12): 3099–3111. doi:10.1093/jxb/erm130. ISSN 0022-0957. PMID 17630292.
70. Managing Photoperiodic Lighting
71. http://flor.hrt.msu.edu/assets/Uploads/Controllingphotoperiod.pdf
72. http://www.hrt.msu.edu/energy/Notebook/pdf/Sec1/Runkle%20and%20Blanchard,%20Use_of_Lighting.pdf
73. Hall, David O.; Rao, Krishna (1999-06-24). Photosynthesis. Cambridge University Press. pp. 8–9. ISBN 9780521644976.
74. Growing your profits: horticultural lighting. Archived August 24, 2013, at the Wayback Machine. Archived August 24, 2013, at the Wayback Machine. Philips (2010). Archived August 24, 2013, at the Wayback Machine.

CHAPTER 3

Horticultural Building System

Chapter Contents

- **Introduction**
- **Horticulture: An Overview**
- **Gardening**
- **Landscaping**

Horticultural Building Systems are defined here as the instance where vegetation and an architectural system exist in a mutually defined and intentionally designed relationship that supports plant growth and an architectonic concept. This definition allows for the history of Horticultural Building Systems to be traced through the seemingly disparate evolutions of horticultural and architectural technology that link the glass house and Crystal Palace to modern architecture and current trends in green architecture. Theories of tectonic culture, modern architecture, and horticultural innovation are placed in direct dialogue with patents and architectural case studies to elucidate a history of Horticultural Building Systems that is inclusive of tectonic, technologic, typological and horticultural histories. Technological and architectural precedents for these system will be presented along with patented technology dating back to 1937.

INTRODUCTION

Horticultural Building Systems are defined as the instance where vegetation and an architectural/architectonic system exist in a mutually defined and intentionally designed relationship that supports plant growth and an architectonic concept. The most common form of these systems in contemporary vernacular is green wall, vertical garden, green roof, roof garden, building-integrated agriculture (BIA), yet the history of these systems may be traced back through greenhouse technology, hydroponicums, horticultural growth chambers, and beyond. These horticultural building systems evolved form a reciprocal relationship between plant cultural requirements and architectural technology.

HORTICULTURE: AN OVERVIEW

Horticulture is the branch of agriculture that deals with the art, science, technology, and business of growing plants. It includes the cultivation of medicinal plants, fruits, vegetables, nuts, seeds, herbs, sprouts, mushrooms, algae, flowers, seaweeds and non-food crops such as grass and ornamental trees and plants. It also includes plant conservation, landscape restoration, landscape and garden design, construction, and maintenance, and arboriculture. Inside agriculture, horticulture contrasts with extensive field farming as well as animal husbandry.

Horticulturists apply their knowledge, skills, and technologies used to grow intensively produced plants for human food and non-food uses and for personal or social needs. Their work involves plant propagation and cultivation with the aim of improving plant growth, yields, quality, nutritional value, and resistance to insects, diseases, and environmental stresses. They work as gardeners, growers, therapists, designers, and technical advisors in the food and non-food sectors of horticulture. Horticulture even refers to the growing of plants in a field or garden.

A fruit and vegetable market in Milan

A vertical garden

Horticultural and health-related produce

Etymology

The word horticulture is modeled after agriculture, and comes from the Greek χόρτος, which in Latin became hortus "garden" and cultūra "cultivation", from cultus, the perfect passive participle of the verb colō "I cultivate". Hortus is cognate with the native English word yard (in the meaning of land associated with a building) and also the borrowed word garden.

Scope

Horticulture involves nine areas of study, which can be grouped into two broad sections: ornamentals and edibles:

- Arboriculture is the study of, and the selection, plant, care, and removal of, individual trees, shrubs, vines, and other perennial woody plants.
- Turf management includes all aspects of the production and maintenance of turf grass for sports, leisure use or amenity use.
- Floriculture includes the production and marketing of floral crops.
- Landscape horticulture includes the production, marketing and maintenance of landscape plants.
- Olericulture includes the production and marketing of vegetables.
- Pomology includes the production and marketing of pome fruits.
- Viticulture includes the production and marketing of grapes.
- Oenology includes all aspects of wine and winemaking.
- Postharvest physiology involves maintaining the quality of and preventing the spoilage of plants and animals.

Anthropology

Horticulture has a very long history. The study and science of horticulture dates all the way back to the times of Cyrus the Great of ancient Persia, and has been going on ever since, with present-day horticulturists such as Freeman S. Howlett and Luther Burbank. The practice of horticulture can be retraced for many thousands of years. The cultivation of taro and yam in Papua New Guinea dates back to at least 6950–6440 cal BP.The origins of horticulture lie in the transition of human communities from nomadic hunter-gatherers to sedentary or semi-sedentary horticultural communities, cultivating a variety of crops on a small scale around their dwellings or in specialized plots visited occasionally during migrations from one area to the next (such as the "milpa" or maize field of Mesoamerican cultures). In the Pre-Columbian Amazon Rainforest, natives are believed to have used biochar to enhance soil productivity by smoldering plant waste. European settlers called it Terra Preta de Indio. In forest areas such horticulture is often carried out in swiddens ("slash and burn" areas). A characteristic of horticultural communities is that useful trees are often to be found planted around communities or specially retained from the natural ecosystem.

Horticulture primarily differs from agriculture in two ways. First, it generally encompasses a smaller scale of cultivation, using small plots of mixed crops rather than large fields of single crops. Secondly, horticultural cultivations generally include a wide variety of crops, even including fruit trees with ground crops. Agricultural cultivations however as a rule focus on one primary crop. In pre-contact North America the semi-sedentary horticultural communities of the Eastern Woodlands (growing maize, squash and sunflower) contrasted markedly with the mobile hunter-gatherer communities of the Plains people. In Central America, Maya horticulture involved augmentation of the forest with useful trees such as papaya, avocado, cacao, ceiba and sapodilla. In the cornfields, multiple crops were grown such as beans (using cornstalks as supports), squash, pumpkins and chilli peppers, in some cultures tended mainly or exclusively by women.

Organizations

The oldest Horticultural society in the world, founded in 1768, is the Ancient Society of York Florists. They still have four shows a year in York, UK.They have a large archive of records, including the original members book dating back to 1768.

Desh Bhagat School of Horticulture Sciences at Desh Bhagat University, Punjab is a prestigious institution in India to encourage the research and promote horticulture techniques throughout the country. The Royal Horticultural Society is a UK charity which oversees several major shows and gardens.

The professional body representing horticulturists in Great Britain and Ireland is the Institute of Horticulture (IOH). Also, the IOH has an international branch for members outside of these islands.

The International Society for Horticultural Science promotes and encourages research and education in all branches of horticultural science.

The American Society of Horticultural Science promotes and encourages research and education in all branches of horticultural science in the Americas.

The Australian Society of Horticultural Science was established in 1990 as a professional society for the promotion and enhancement of Australian horticultural science and industry.

The National Junior Horticultural Association (NJHA) was established in 1934 and was the first organization in the world dedicated solely to youth and horticulture. NJHA programs are designed to help young people obtain a basic understanding of, and develop skills in, the ever expanding art and science of horticulture.

The New Zealand Horticulture Institute.

The Global Horticulture Initiative (GlobalHort) fosters more efficient and effective partnerships and collective action among different stakeholders in horticulture. The organization has a special focus on horticulture for development (H4D), i.e. using horticulture to reduce poverty and improve nutrition worldwide. To be efficient, GlobalHort is organized in a consortium of national and international organizations to collaborate in research, training, and technology-generating activities designed to meet mutually-agreed-upon objectives. GlobalHort is a not-for-profit organization registered in Belgium.

GARDENING

Part of a parterre in an English garden

Gardening is the practice of growing and cultivating plants as part of horticulture. In gardens, ornamental plants are often grown for their flowers, foliage, or overall appearance; useful plants, such as root vegetables, leaf vegetables, fruits, and herbs, are grown for consumption, for use as dyes, or for medicinal or cosmetic use. Gardening is considered to be a relaxing activity for many people.

Gardening ranges in scale from fruit orchards, to long boulevard plantings with one or more different types of shrubs, trees, and herbaceous plants, to residential yards including lawns and foundation plantings, to plants in large or small containers grown inside or outside. Gardening may be very specialized, with only one type of plant grown, or involve a large number of different plants in mixed plantings. It involves an active participation in the growing of plants, and tends to be labor-intensive, which differentiates it from farming or forestry.

Ancient times

Forest gardening, a forest-based food production system, is the world's oldest form of gardening. Forest gardens originated in prehistoric times along jungle-clad river banks and in the wet foothills of monsoon regions. In the gradual process of families improving their immediate environment, useful tree and vine species were identified, protected and improved while undesirable species were eliminated. Eventually foreign species were also selected and incorporated into the gardens.

After the emergence of the first civilizations, wealthy individuals began to create gardens for aesthetic purposes. Egyptian tomb paintings from around 1500 BC provide some of the earliest physical evidence of ornamental horticulture and landscape design; they depict lotus ponds surrounded by symmetrical rows of acacias and palms. A notable example of ancient ornamental gardens were the Hanging Gardens of Babylon—one of the Seven Wonders of the Ancient World —while ancient Rome had dozens of gardens.

Wealthy ancient Egyptians used gardens for providing shade. Egyptians associated trees and gardens with gods as they believed that their deities were pleased by gardens. Gardens in ancient Egypt were often surrounded by walls with trees planted in rows. Among the most popular species planted were date palms, sycamores, fir trees, nut trees, and willows. These gardens were a sign of higher socioeconomic status. In addition, wealthy ancient Egyptians grew vineyards, as wine was a sign of the higher social classes. Roses, poppies, daisies and irises could all also be found in the gardens of the Egyptians.

Assyria was also renowned for its beautiful gardens. These tended to be wide and large, some of them used for hunting game—rather like a game reserve today—and others as leisure gardens. Cypresses and palms were some of the most frequently planted types of trees.

Ancient Roman gardens were laid out with hedges and vines and contained a wide variety of flowers—acanthus, cornflowers, crocus, cyclamen, hyacinth, iris, ivy, lavender, lilies, myrtle, narcissus, poppy, rosemary and violets—as well as statues and sculptures. Flower beds were popular in the courtyards of rich Romans.

A gardener at work, 1607

The middle Ages

The Middle Age represented a period of decline in gardens for aesthetic purposes, with regard to gardening. After the fall of Rome, gardening was done for the purpose of growing medicinal herbs and/or decorating church altars. Monasteries carried on a tradition of garden design and intense horticultural techniques during the medieval period in Europe. Generally, monastic garden types consisted of kitchen gardens, infirmary gardens, cemetery orchards, cloister garths and vineyards. Individual monasteries might also have had a "green court", a plot of grass and trees where horses could graze, as well as a cellarer's garden or private gardens for obedientiaries, monks who held specific posts within the monastery.

Islamic gardens were built after the model of Persian gardens and they were usually enclosed by walls and divided in 4 by watercourses. Commonly, the center of the garden would have a pool or pavilion. Specific to the Islamic gardens are the mosaics and glazed tiles used to decorate the rills and fountains that were built in these gardens.

By the late 13th century, rich Europeans began to grow gardens for leisure and for medicinal herbs and vegetables. They surrounded the gardens by walls to protect them from animals and to provide seclusion. During the next two centuries, Europeans started planting lawns and raising flowerbeds and trellises of roses. Fruit trees were common in these gardens and also in some, there were turf seats. At the same time, the gardens in the monasteries were a place to grow flowers and medicinal herbs but they were also a space where the monks could enjoy nature and relax.

The gardens in the 16th and 17th century were symmetric, proportioned and balanced with a more classical appearance. Most of these gardens were built around a central axis and they were divided into different parts by hedges. Commonly, gardens had flowerbeds laid out in squares and separated by gravel paths.

Gardens in Renaissance were adorned with sculptures, topiary and fountains. In the 17th century, knot gardens became popular along with the hedge mazes. By this time, Europeans started planting new flowers such as tulips, marigolds and sunflowers.

Cottage Gardens

Cottage gardens, which emerged in Elizabethan times, appear to have originated as a local source for herbs and fruits. One theory is that they arose out of the Black Death of the 1340s, when the death of so many laborers made land available for small cottages with personal gardens.According to the late 19th-century legend of origin, these gardens were originally created by the workers that lived in the cottages of the villages, to provide them with food and herbs, with flowers planted among them for decoration. Farm workers were provided with cottages that had architectural quality set in a small garden—about an acre—where they could grow food and keep pigs and chickens.

Authentic gardens of the yeoman cottager would have included a beehive and livestock, and frequently a pig and sty, along with a well. The peasant cottager of medieval times was more interested in meat than flowers, with herbs grown for medicinal use rather than for their beauty. By Elizabethan times there was more prosperity, and thus more room to grow flowers. Even the early cottage garden flowers typically had their practical use—violets were spread on the floor (for their pleasant scent and keeping out vermin); calendulas and primroses were both attractive and used in cooking. Others, such as sweet william and hollyhocks, were grown entirely for their beauty.

18th Century

In the 18th century, gardens were laid out more naturally, without any walls. This style of smooth undulating grass, which would run straight to the house, clumps, belts and scattering of trees and his serpentine lakes formed by invisibly damming small rivers, were a new style within the English landscape, a "gardenless" form of landscape gardening, which swept away almost all the remnants of previous formally patterned styles. The English garden usually included a lake, lawns set against groves of trees, and often contained shrubberies, grottoes, pavilions, bridges and follies such as mock temples, Gothic ruins, bridges, and other picturesque architecture, designed to recreate an idyllic pastoral landscape. This new style emerged in England in the early 18th century, and spread across Europe, replacing the more formal, symmetrical Garden à la française of the 17th century as the principal gardening style of Europe. The English garden presented an idealized view of nature. They were often inspired by paintings of landscapes by Claude Lorraine and Nicolas Poussin, and some were Influenced by the classic Chinese gardens of the East, which had recently been described by European travelers. The work of Lancelot ‹Capability› Brown was particularly influential. Also, in 1804 the Horticultural Society was formed. Gardens of the 19th century contained plants such as the monkey puzzle or Chile pine. This is also the time when the so-called gardenesque style of gardens evolved. These gardens displayed a wide variety of flowers in a rather small space. Rock gardens increased in popularity in the 19th century.

Types

Conservatory of Flowers in Golden Gate Park, San Francisco

Hanging baskets in Thornbury, South Gloucestershire

Residential gardening takes place near the home, in a space referred to as the **garden**. Although a garden typically is located on the land near a residence, it may also be located on a roof, in an atrium, on a balcony, in a windowbox, or on a patio or vivarium.

Gardening also takes place in non-residential green areas, such as parks, public or semi-public gardens (botanical gardens or zoological gardens), amusement and amusement parks, along transportation corridors, and around tourist attractions and garden hotels. In these situations, a staff of gardeners or groundskeepers maintains the gardens.

- Indoor gardening is concerned with the growing of houseplants within a residence or building, in a conservatory, or in a greenhouse. Indoor gardens are sometimes incorporated as part of air conditioning or heating systems. Indoor gardening extends the growing season in the fall and spring and can be used for winter gardening.
- Native plant gardening is concerned with the use of native plants with or without the intent of creating wildlife habitat. The goal is to create a garden in harmony with, and adapted to a given area. This type of gardening typically reduces water usage, maintenance, and fertilization costs, while increasing native faunal interest.
- Water gardening is concerned with growing plants adapted to pools and ponds. Bog gardens are also considered a type of water garden. These all require special conditions and considerations. A simple water garden may consist solely of a tub containing the water and plant(s). In aquascaping, a garden is created within an aquarium tank.
- Container gardening is concerned with growing plants in any type of container either indoors or outdoors. Common containers are pots, hanging baskets, and planters. Container gardening is usually used in atriums and on balconies, patios, and roof tops.

- Hügelkultur is concerned with growing plants on piles of rotting wood, as a form of raised bed gardening and composting in situ. An English loanword from German, it means "mound garden." Toby Hemenway, noted Permaculture author and teacher, considers wood buried in trenches to also be a form of hugelkultur referred to as a dead wood swale. Hugelkultur is practiced by Sepp Holzer as a method of forest gardening and agroforestry, and by Geoff Lawton as a method of dryland farming and desert greening. When used as a method of disposing of large volumes of waste wood and woody debris, hugelkultur accomplishes carbon sequestration. It is also a form of xeriscaping.
- Community gardening is a social activity in which an area of land is gardened by a group of people, providing access to fresh produce and plants as well as access to satisfying labor, neighborhood improvement, sense of community and connection to the environment.Community gardens are typically owned in trust by local governments or nonprofits.
- Garden sharing partners landowners with gardeners in need of land. These shared gardens, typically front or back yards, are usually used to produce food that is divided between the two parties.
- Organic gardening uses natural, sustainable methods, fertilizers and pesticides to grow non-genetically modified crops.

Garden Features and Accessories

There is a wide range of features and accessories available in the market for both the professional gardener and the amateur to exercise their creativity. These are used to add decoration or functionality, and may be made from a wide range of materials such as copper, stone, wood, bamboo, stainless steel, clay, stained glass, concrete, or iron. Examples include trellis, arbors, statues, benches, water fountains, urns, bird baths and feeders, and garden lighting such as candle lanterns and oil lamps. The use of these items can be part of the expression of a gardener's gardening personality.

Gardening Departments and Centers

Gardening departments and centers mainly sell plants, sundries, and garden accessories, but in recent times, many now stock outdoor leisure products as diverse as spas, furniture, and barbecues. Many garden centers now include food halls, and sections for clothing, gifts, pets, and power tools. There are also a number of online garden centers that now deliver directly to customers' doors.

Comparison with Farming

Hand gardening tools

Gardening for beauty is likely nearly as old as farming for food, however for most of history for the majority of people there was no real distinction since the need for food and other useful product trumped other concerns. Small-scale, subsistence agriculture (called hoe-farming) is largely indistinguishable from gardening. A patch of potatoes grown by a Peruvian peasant or an Irish smallholder for personal use could be described as either a garden or a farm. Gardening for average people evolved as a separate discipline, more concerned with aesthetics and recreation, under the influence of the pleasure gardens of the wealthy. Meanwhile, farming has evolved (in developed countries) in the direction of commercialization, economics of scale, and monocropping.

In respect to its food producing purpose, gardening is distinguished from farming chiefly by scale and intent. Farming occurs on a larger scale, and with the production of salable goods as a major motivation. Gardening is done on a smaller scale, primarily for pleasure and to produce goods for the gardener's own family or community. There is some overlap between the terms, particularly in that some moderate-sized vegetable growing concerns, often called market gardening, can fit in either category.

The key distinction between gardening and farming is essentially one of scale; gardening can be a hobby or an income supplement, but farming is generally understood as a full-time or commercial activity, usually involving more land and quite different practices. One distinction is that gardening is labor-intensive and employs very little infrastructural capital, sometimes no more than a few tools, e.g. a spade, hoe, basket and watering can. By contrast, larger-scale farming often involves irrigation systems, chemical fertilizers and harvesters or at least ladders, e.g. to reach up into fruit trees. However, this distinction is becoming blurred with the increasing use of power tools in even small gardens.

In part because of labor intensity and aesthetic motivations, gardening is very often much more productive per unit of land than farming. In the Soviet Union, half the food supply came from small peasants' garden plots on the huge government-run collective farms, although they were tiny patches of land. Some argue this as evidence of the superiority of capitalism, since the peasants were generally able to sell their produce. Others consider it to be evidence of a tragedy of the commons, since the large collective plots were often neglected, or fertilizers or water redirected to the private gardens.

Planting in a garden

The term precision agriculture is sometimes used to describe gardening using intermediate technology (more than tools, less than harvesters), especially of organic varieties. Gardening is effectively scaled up to feed entire villages of over 100 people from specialized plots. A variant is the community garden which offers plots to urban dwellers; see further in allotment (gardening).

Gardens as Art

Garden at the Schultenhof in Mettingen, North Rhine-Westphalia, Germany

Garden design is considered to be an art in most cultures, distinguished from gardening, which generally means garden maintenance. Garden design can include different themes such as perennial, butterfly, wildlife, Japanese, water, tropical, or shade gardens. In Japan, Samurai and Zen monks were often required to build decorative gardens or practice related skills like flower arrangement known as ikebana. In 18th-century Europe, country estates were refashioned by landscape gardeners into formal gardens or landscaped park lands, such as at Versailles, France, or Stowe, England. Today, landscape architects and garden designers continue to produce artistically creative designs for private garden spaces. In the USA, professional landscape designers are certified by the Association of Professional Landscape Designers.

Social Aspects

People can express their political or social views in gardens, intentionally or not. The lawn vs. garden issue is played out in urban planning as the debate over the "land ethic" that is to determine urban land use and whether hyper hygienist bylaws (e.g. weed control) should apply, or whether land should generally be allowed to exist in its natural wild state. In a famous Canadian Charter of Rights case, "Sandra Bell vs. City of Toronto", 1997, the right to cultivate all native species, even most varieties deemed noxious or allergenic, was upheld as part of the right of free expression.

Community gardening comprises a wide variety of approaches to sharing land and gardens.

People often surround their house and garden with a hedge. Common hedge plants are privet, hawthorn, beech, yew, leyland cypress, hemlock, arborvitae, barberry, box, holly, oleander, forsythia and lavender. The idea of open gardens without hedges may be distasteful to those who enjoy privacy. The Slow Food movement has sought in some countries to add an edible school yard and garden classrooms to schools, e.g. in Fergus, Ontario, where these were added to a public school to augment the kitchen classroom. Garden sharing, where urban landowners allow gardeners to grow on their property in exchange for a share of the harvest, is associated with the desire to control the quality of one's food, and reconnect with soil and community.

In US and British usage, the production of ornamental plantings around buildings is called landscaping, landscape maintenance or grounds keeping, while international usage uses the term gardening for these same activities.

Also gaining popularity is the concept of Green Gardening which involves growing plants using organic fertilizers and pesticides so that the gardening process - or the flowers and fruits produced thereby - doesn›t adversely affect the environment or people›s health in any manner.

Garden Pests

Garden pests are generally plants, fungi, or animals (frequently insects) that engages in activity that the gardener considers undesirable. It may crowd out desirable plants, disturb soil, stunt the growth of young seedlings, steal or damage fruit, or otherwise kill plants, hamper their growth, damage their appearance, or reduce the quality of the edible or ornamental portions of the plant. Aphids, spider mites, slugs, snails, ants, birds, and even cats are commonly considered to be garden pests.

Because gardeners may have different goals, organisms considered "garden pests" vary from gardener to gardener. Tropaeolum speciosum, for example, may be considered a desirable and ornamental garden plant, or it may be considered a pest if it seeds and starts to grow where it is not wanted. As another example, in lawns, moss can become dominant and be impossible to eradicate. In some lawns, lichens, especially very damp lawn lichens such as Peltigera lactucfolia and P. membranacea, can become difficult to control and be considered pests.

Garden Pest Control

There are many ways by which unwanted pests are removed from a garden. The techniques vary depending on the pest, the gardener's goals, and the gardener's philosophy. For example, snails may be dealt with through the use of a chemical pesticide, an organic pesticide, hand-picking, barriers, or simply growing snail-resistant plants.

Pest control is often done through the use of pesticides, which may be either organic or artificially synthesized. Pesticides may affect the ecology of a garden due to their effects on the populations of both target and non-target species. For example, unintended exposure to some neonicotinoid pesticides has been proposed as a factor in the recent decline in honey bee populations. A mole vibrator can deter mole activity in a garden.

Other means of control include the removal of infected plants, using fertilizers and biostimulants to improve the health and vigour of plants so they better resist attack, practising crop rotation to prevent pest build-up, using companion planting, and practising good garden hygiene, such as disinfecting tools and clearing debris and weeds which may harbour pests.

VERTICAL FARMING

Vertical farming is the practice of producing food in vertically stacked layers, such as in a skyscraper, used warehouse, or shipping container. The modern ideas of vertical farming use indoor farming techniques and controlled-environment agriculture (CEA) technology, where all environmental factors can be controlled. These facilities utilize artificial control of light, environmental control (humidity, temperature, gases...) and fertigation. Some vertical farms use techniques similar to greenhouses, where natural sunlight can be augmented with artificial lighting and metal reflectors.

Lettuce grown in indoor vertical farming system.

Types

The term "vertical farming" was coined by Gilbert Ellis Bailey in 1915 in his book Vertical Farming. His use of the term differs from the current meaning—he wrote about farming with a special interest in soil origin, its nutrient content and the view of plant life as "vertical" life forms, specifically relating to their underground root structures. Modern usage of the term "vertical farming" usually refers to growing plants in layers, whether in a multistory skyscraper, used warehouse, or shipping container.

Mixed-Use Skyscrapers

Mixed-use skyscrapers were proposed and built by architect Ken Yeang. Yeang proposes that instead of hermetically sealed mass-produced agriculture, plant life should be cultivated within open air, mixed-use skyscrapers for climate control and consumption. This version of vertical farming is based upon personal or community use rather than the wholesale production and distribution plant life that aspires to feed an entire city. It thus requires less of an initial investment than Despommier's "vertical farm". However, neither Despommier nor Yeang are the conceptual originators, nor is Yeang the inventor of vertical farming in skyscrapers.

Despommier's Skyscrapers

Ecologist Dickson Despommier argues that vertical farming is legitimate for environmental reasons. He claims that the cultivation of plant life within skyscrapers will require less embedded energy and produce less pollution than some methods of producing plant life on natural landscapes. He moreover claims that natural landscapes are too toxic for natural, agricultural production, despite the ecological and environmental costs of extracting materials to build skyscrapers for the simple purpose of agricultural production.

Dickson Despommier shares his ideas about how "vertical farming" can help reduce hunger by changing the way we use land for agriculture. Photography by Kris Krüg

Vertical farming according to Despommier thus discounts the value of natural landscape in exchange for the idea of skyscraper as spaceship. Plant life is mass-produced within hermetically sealed, artificial environments that have little to do with the outside world. In this sense, they could be built anywhere regardless of the context. Although climate control, lighting, and other costs of maintenance have been posited as potentially stifling to bringing this concept to fruition, advocates have countered that an important feature of future vertical farms will be the integration of renewable energy technology, be it solar panels, wind turbines, water capture systems, and probably some combination of the three. The vertical farm is designed to be sustainable, and to enable nearby inhabitants to work at the farm.

Despommier›s concept of the vertical farm emerged in 1999 at Columbia University. It promotes the mass cultivation of plant life for commercial purposes in skyscrapers.

Stackable Shipping Containers

Several companies have brought forth the concept of stacking recycled shipping containers in urban settings. Freight Farms produces a "leafy green machine" that is a complete farm-to-table system outfitted with vertical hydroponics, LED lighting and intuitive climate controls built within a 12 m × 2.4 m shipping container. Podponics has built a large scale vertical farm in Atlanta consisting of over 100 stacked "growpods". A similar farm is currently under construction in Oman.

Problems

Economics

Opponents question the potential profitability of vertical farming. A detailed cost analysis of start-up costs, operation costs and revenue has not been done. The extra cost of lighting, heating, and powering the vertical farm may negate any of the cost benefits received by the decrease in transportation expenses. The economic and environmental benefits of vertical farming rest partly on the concept of minimizing food miles, the distance that food travels from farm to consumer. However, a recent analysis suggests that transportation is only a minor contributor to the economic and environmental costs of supplying food to urban populations. The author of the report, University of Toronto professor Pierre Desrochers, concluded that "food miles are, at best, a marketing fad." Thus the facility would have to produce a considerable profit to justify remaining in the city. A simpler concept rather than trying to stack farms on multiple stories would be to just cultivate crops on the roofs of existing building. Rooftop farming is a growing urban trend, requires little construction (other than fortifying the roof to hold the weight of the growing medium), still takes advantage of sunlight and doesn't require investment in machinery, growing lights or irrigation.

Similarly, if the power needs of the vertical farm are met by fossil fuels, the environmental effect may be a net loss; even building low-carbon capacity to power the farms may not make as much sense as simply leaving the traditional farms in place, and burning less coal.

The initial building costs will be easily over $100 million, for a 60 hectare vertical farm. Office occupancy costs can be very high in major cities, with cities such as Tokyo, Moscow, Mumbai, Dubai, Milan, Zurich, and Sao Paulo ranging from $1850 to $880 per square meter, respectively.

Energy Use

During the growing season, the sun shines on a vertical surface at an extreme angle such that much less light is available to crops than when they are planted on flat land. Therefore, supplemental light would be required in order to obtain economically viable yields. Bruce Bugbee, a crop physiologist at Utah State University, believes that the power demands of vertical farming will be too expensive and uncompetitive with traditional farms using only free natural light. The environmental writer George Monbiot calculated that the cost of providing enough supplementary light to grow the grain for a single loaf would be about $15. An article in the Economist argued that "even though crops growing in a glass skyscraper will get some natural sunlight during the day, it won't be enough" and "the cost of powering artificial lights will make indoor farming prohibitively expensive".

As "The Vertical Farm" proposes a controlled environment, heating and cooling costs will be at least as costly as any other tower. But there also remains the issue of complicated, if not more expensive, plumbing and elevator systems to distribute food and water throughout. Even throughout the northern continental United States, while heating with relatively cheap fossil fuels, the heating cost can be over $200,000 per hectare.

To address this problem, The Plant in Chicago is building an anaerobic digester into the building. This will allow the farm to operate off the energy grid. Moreover, the anaerobic digester will be recycling waste from nearby businesses that would otherwise go into landfills.

Pollution

Depending on the method of electricity generation used, regular greenhouse produce can create more greenhouse gases than field produce,largely due to higher energy use per kilogram of produce. With vertical farms requiring much greater energy per kilogram of produce, mainly through increased lighting, than regular greenhouses, the amount of pollution created will be much higher than that from field produce. The amount of pollution produced is dependent on how the energy used in the process is generated.

As plants acquire nearly all their carbon from the atmosphere, greenhouse growers commonly supplement CO_2 levels to 3–4 times the rate normally found in the atmosphere. This increase in CO_2, which has been shown to increase photosynthesis rates by 50%, contributes to the higher yields expected in vertical farming. It is not uncommon to find greenhouses burning fossil fuels purely for this purpose, as other CO_2 sources, like from furnaces, contain pollutants such as sulphur dioxide and ethylene which significantly damage plants. This means a vertical farm will require a CO_2 source, most likely from combustion, even if the rest of the farm is powered by "green" energy. Also, through necessary ventilation, much CO_2 will be leaked into the city's atmosphere.

Greenhouse growers commonly exploit photoperiodism in plants to control whether the plants are in a vegetative or reproductive stage. As part of this control, growers will have the lights on past sunset and before sunrise or periodically throughout the night. Single story greenhouses are already a nuisance to neighbours because of light pollution, a 30-story vertical farm in a densely populated area will surely face problems because of its light pollution.

Hydroponics greenhouses regularly change the water, meaning there is a large quantity of water containing fertilizers and pesticides that must be disposed of. While solutions are currently being worked on, the most common method of simply spreading the mixture over a sufficient area of neighbouring farmland or wetlands would be more difficult for an urban vertical farm.

Advantages

Several potential advantages of vertical farming have been discussed by Despommier. Many of these benefits are obtained from scaling up hydroponic or aeroponic growing methods.

Preparation for the Future

It is estimated that by the year 2050, close to 80% of the world's population will live in urban areas and the total population of the world will increase by 3 billion people. A very large amount of land may be required depending on the change in yield per hectare.

Scientists are concerned that this large amount of required farmland will not be available and that severe damage to the earth will be caused by the added farmland. According to Despommier, vertical farms, if designed properly, may eliminate the need to create additional farmland and help create a cleaner environment.

Increased Crop Production

Unlike traditional farming in non-tropical areas, indoor farming can produce crops year-round. All-season farming multiplies the productivity of the farmed surface by a factor of 4 to 6 depending on the crop. With some crops, such as strawberries, the factor may be as high as 30.

Furthermore, as the crops would be sold in the same infrastructures in which they are grown, they will not need to be transported between production and sale, resulting in less spoilage, infestation, and energy required than conventional farming encounters. Research has shown that 30% of harvested crops are wasted due to spoilage and infestation, though this number is much lower in developed nations.

Despommier suggests that, if dwarf versions of certain crops are used (e.g. dwarf wheat, which has been grown in space by NASA, is smaller in size but richer in nutrients), year-round crops, and "stacker" plant holders are accounted for, a 30-story building with a base of a building block (2 hectares (5 acres)) would yield a yearly crop analogous to that of 1,000 hectares (2,400 acres) of traditional farming.

Protection from Weather-Related Problems

Crops grown in traditional outdoor farming suffer from the often suboptimal, and sometimes extreme, nature of geological and meteorological events such as undesirable temperatures or rainfall amounts, monsoons, hailstorms, tornadoes, flooding, wildfires, and severe droughts. The protection of crops from weather is increasingly important as global climate change occurs. "Three recent floods (in 1993, 2007 and 2008) cost the United States billions of dollars in lost crops, with even more devastating losses in topsoil. Changes in rain patterns and temperature could diminish India's agricultural output by 30 percent by the end of the century."

Because vertical plant farming provides a controlled environment, the productivity of vertical farms would be mostly independent of weather and protected from extreme weather events. Although the controlled environment of vertical farming negates most of these factors, earthquakes and tornadoes still pose threats to the proposed infrastructure, although this again depends on the location of the vertical farms.

Conservation of Resources

Each unit of area in a vertical farm could allow up to 20 units of area of outdoor farmland to return to its natural state, and recover farmlands due to development from original flat farmlands.

Vertical farming would reduce the need for new farmland due to overpopulation, thus saving many natural resources, currently threatened by deforestation or pollution.

Deforestation and desertification caused by agricultural encroachment on natural biomes would be avoided. Because vertical farming lets crops be grown closer to consumers, it would substantially reduce the amount of fossil fuels currently used to transport and refrigerate farm produce. Producing food indoors reduces or eliminates conventional plowing, planting, and harvesting by farm machinery, also powered by fossil fuels.

Halting Mass Extinction

Withdrawing human activity from large areas of the Earth's land surface may be necessary to slow and eventually halt the current anthropogenic mass extinction of land animals.

Traditional agriculture is highly disruptive to wild animal populations that live in and around farmland and some argue it becomes unethical when there is a viable alternative. One study showed that wood mouse populations dropped from 25 per hectare to 5 per hectare after harvest, estimating 10 animals killed per hectare each year with conventional farming. In comparison, vertical farming would cause very little harm to wildlife, and would allow disused farmland to return to its pre-agricultural state.

Impact on Human Health

Traditional farming is a hazardous occupation with particular risks that often take their toll on the health of human laborers. Such risks include: exposure to infectious diseases such as malaria and schistosomes, exposure to toxic chemicals commonly used as pesticides and fungicides, confrontations with dangerous wildlife such as venomous snakes, and the severe injuries that can occur when using large industrial farming equipment. Whereas the traditional farming environment inevitably contains these risks (particularly in the farming practice known as "slash and burn"), vertical farming – because the environment is strictly controlled and predictable – reduces some of these dangers. Currently, the American food system makes fast, unhealthy food cheap while fresh produce is less available and more expensive, encouraging poor eating habits. These poor eating habits lead to health problems such as obesity, heart disease, and diabetes. The increased availability and subsequent lower cost of fresh produce would encourage healthier eating.

Poverty/Destitution and Culture

Food security is one of the primary factors leading to absolute poverty. Being able to construct 'farm land' in secure areas as needed will help alleviate the pressures causing crises among neighbors fighting for resources (mainly water and space). It also allows continued growth of culturally significant food items without sacrificing sustainability or basic needs, which can be significant to the recovery of a society from poverty.

Urban Growth

Vertical farming, used in conjunction with other technologies and socioeconomic practices, could allow cities to expand while remaining largely self-sufficient food wise. This would allow for large urban centers that could grow without destroying considerably larger areas of forest to provide food for their people. Moreover, the industry of vertical farming will provide employment to these expanding urban centers. This may help displace the unemployment created by the dismantling of traditional farms, as more farm laborers move to cities in search of work.

Energy Sustainability

Vertical farms could exploit methane digesters to generate a small portion of its own electrical needs. Methane digesters could be built on site to transform the organic waste generated at the farm into biogas which is generally composed of 65% methane along with other gases. This biogas could then be burned to generate electricity for the greenhouse.

Technologies and Devices

Vertical farming relies on the use of various physical methods to become effective. Combining these technologies and devices in an integrated whole is necessary to make Vertical Farming a reality. Various methods are proposed and under research. The most common technologies suggested are:

- Greenhouse
- The Folkewall and other vertical growing architectures
- Flowerpot
- Aeroponics / Hydroponics / Aquaponics
- Composting
- Grow light
- Phytoremediation
- Skyscraper
- Controlled-environment agriculture
- Precision agriculture
- Agricultural robot

Plans

Despommier argues that the technology to construct vertical farms currently exists. He also states that the system can be profitable and effective, a claim evidenced by some preliminary research posted on the project's website. Developers and local governments in the following cities have expressed serious interest in establishing a vertical farm: Incheon (South Korea), Abu Dhabi (United Arab Emirates), and Dongtan (China), New York City, Portland, Ore., Los Angeles, Las Vegas, Seattle, Surrey, B.C., Toronto, Paris, Bangalore, Dubai, Shanghai and Beijing. The Illinois Institute of Technology is now crafting a detailed plan for Chicago. It is suggested that prototype versions of vertical farms should be created first, possibly at large universities interested in the research of vertical farms, in order to prevent failures such as the Biosphere 2 project in Oracle,

Arizona.

In 2009, the world's first pilot production system was installed at Paignton Zoo Environmental Park in the United Kingdom. The project showcased a technological solution for vertical farming and provided a physical base to conduct research into sustainable urban food production. The produce is used to feed the zoo's animals while the project enables evaluation of the systems and provides an educational resource to advocate for change in unsustainable land use practices that impact upon global biodiversity and ecosystem services,

In 2010, the Green Zionist Alliance proposed a resolution at the 36th World Zionist Congress calling on Keren Kayemet L'Yisrael (Jewish National Fund in Israel) to develop vertical farms in Israel.

In 2012, the world's first commercial vertical farm was opened in Singapore, developed by Sky Greens Farms, and is three stories high. They currently have over 100 towers that stand at nine meters tall.

In 2013 (July 18) the Association for Vertical Farming (AVF) was founded in Munich (Germany). By May 2015 the AVF already expanded with regional chapters all over Europe, Asia, USA, Canada and the United Kingdom. This internationally active non-profit organization unites growers and inventors to improve food security and the sustainable development of Vertical Farming. The AVF also focuses on advancing vertical farming technologies, designs and businesses by hosting international info-days, workshops and summits. They developed a glossary to bring consistency to the industry and plan on helping to standardize the technologies.

LANDSCAPING

Landscaping refers to any activity that modifies the visible features of an area of land, including:

- living elements, such as flora or fauna; or what is commonly called gardening, the art and craft of growing plants with a goal of creating a beautiful environment within the landscape.
- natural elements such as landforms, terrain shape and elevation, or bodies of water; and
- abstract elements such as the weather and lighting conditions.

Landscaping requires expertise in horticulture and artistic design.

Understanding the Land

Construction requires study and observation. It is not the same in different parts of the world. Landscaping varies according to different regions.Therefore, normally local natural experts are recommended if it is done for the first time. Understanding of the site is one of the chief essentials for successful landscaping. Different natural features like terrain, topography, soil qualities, prevailing winds, depth of the frost line, and the system of native flora and fauna must be taken into account. Sometimes the land is not fit for landscaping. In order to landscape it, the land must be reshaped. This reshaping of land is called grading.

Removal of earth from the land is called cutting while when earth is added to the slope, it is called filling. Sometimes the grading process may involve removal of excessive waste (landfills), soil and rocks, so designers should take into account while in the planning stage.

Tools

In the start, the landscaping contractor makes a letter which is a rough design and layout of what could be done with the land in order to achieve the desired outcome. Different pencils are required to make graphics of the picture. Landscaping has become more technological than natural, as few projects begin without bulldozers, lawnmowers, or chainsaws. Different areas have different qualities of plants. Fertilizers are required for this purpose in excess amounts as natural landscaping is done. Some landscapers prefer to use mix gravel with rocks of varying sizes to add interest in large areas.

REFERENCES

1. hortus. Charlton T. Lewis and Charles Short. A Latin Dictionary on Perseus Project.
2. Harper, Douglas. "horticulture". Online Etymology Dictionary.
3. Entry for yard Dictionary.com (presenting information supposedly from Random House Dictionary)
4. "Archived copy". Archived from the original on September 10, 2012. Retrieved September 21, 2012.
5. Fullagar, Richard, Judith Field, Tim Denham, and Carol Lentfer (2006) Early and mid Holocene tool-use and processing of taro (Colocasia esculenta), yam (Dioscorea sp.) and other plants at Kuk Swamp in the highlands of Papua New Guinea Journal of Archaeological Science 33: 595–614
6. von Hagen, V.W. (1957) The Ancient Sun Kingdoms Of The Americas. Ohio: The World Publishing Company
7. Solomon, Dawit, Johannes Lehmann, Janice Thies, Thorsten Schafer, Biqing Liang, James Kinyangi, Eduardo Neves, James Petersen, Flavio Luizao, and Jan Skjemstad, Molecular signature and sources of biochemical recalcitrance of organic carbone in Amazonian Dark Earths, Geochemica et cosmochemica ACTA 71.9 2285-2286 (2007) ("Amazonian Dark Earths (ADE) are a unique type of soils apparently developed between 500 and 9000 years B.P. through intense anthropogenic activities such as biomass-burning and high-intensity nutrient depositions on pre-Columbian Amerindian settlements that transformed the original soils into Fimic Anthrosols throughout the Brazilian Amazon Basin.") (internal citations omitted)

8. Glaser, Bruno, Johannes Lehmann, and Wolfgang Zech, Ameliorating physical and chemical properties of highly weathered soils in the tropics with charcoal – a review, Biology and Fertility of Soils 35.4 219-220 (2002) ("These so called Terra Preta do Indio (Terra Preta) characterize the settlements of pre-Columbian Indios. In Terra Preta soils large amounts of black C indicate a high and prolonged input of carbonized organic matter probably due to the production of charcoal in hearths, whereas only low amounts of charcoal are added to soils as a result of forest fires and slash-and-burn techniques.") (internal citations omitted)
9. McGee, J.R. and Kruse, M. (1986) Swidden horticulture among the Lacandon Maya videorecording (29 mins.). University of California, Berkeley: Extension Media Center
10. Thompson, S.I. (1977) Women, Horticulture, and Society in Tropical America. American Anthropologist, N.S., 79: 908–10
11. Douglas John McConnell (2003). The Forest Farms of Kandy: And Other Gardens of Complete Design. p. 1. ISBN 9780754609582.
12. Douglas John McConnell (1992). The forest-garden farms of Kandy, Sri Lanka. p. 1. ISBN 9789251028988.
13. "A Brief History of Gardening". Retrieved 2010-06-04.
14. Ryrie, Charlie (2004). The Cottage Garden: How to Plan and Plant a Garden That Grows Itself. Collins & Brown. p. 7. ISBN 1-84340-216-5.
15. Scott-James, Anne; Osbert Lancaster (2004). The Pleasure Garden: An Illustrated History of British Gardening. Frances Lincoln Publishers. p. 80. ISBN 978-0-7112-2360-8.
16. Anne Scott-James, The Cottage Garden (London: Lane) 1981, de-mythologised the origins of the English cottage garden, and its treasured topiary among the vegetables and flowers, popularly supposed to represent heirlooms from the seventeenth century.
17. A Biographical Dictionary of British Architects, 1600–1840, Howard Colvin, Yale University Press, 2008 ISBN 0-300-12508-9, p 659
18. Lloyd, Christopher; Richard Bird (1999). The Cottage Garden. Jacqui Hurst. Dorling Kindersley. pp. 6–9. ISBN 978-0-7513-0702-3.
19. Yves-Marie Allain and Janine Christiany, L'Art des jardins en Europe, Citadelles and Mazenod, Paris, 2006.
20. Boults, Elizabeth and Chip Sullivan (2010). Illustrated History of Landscape Design. John Wiley and Sons. p. 175. ISBN 0-470-28933-3.
21. Hemenway, Toby (2009). Gaia's Garden: A Guide to Home-Scale Permaculture. Chelsea Green Publishing. pp. 84-85. ISBN 978-1-60358-029-8.
22. http://permaculture.org.au/2009/12/11/greening-the-desert-ii-final/
23. "What is a community garden?". American Community Garden Association. 2007.
24. Hannah, A. K. & Oh, P. (2000) Rethinking Urban Poverty: A look at Community Gardens. Bulletin of Science, Technology and & Society. 20(3). 207-216.

25. Ferris, J., Norman, C. & Sempik, J. (2001) People, Land and Sustainability: Community Gardens and the Social Dimension of Sustainable Development. Social Policy and Administration. 35(5). 559-568.
26. APLD.org
27. Meet the urban sharecroppers The Guardian, 4 September 2008
28. "A Common Pesticide Decreases Foraging Success and Survival in Honey Bees". Retrieved 2012-03-29.
29. "Mole-ested". Retrieved 2014-05-28.
30. "The Self-Sufficient Gardener Podcast--Episode 24 Companion Planting and Crop Rotation". Retrieved 2010-08-13.
31. Hix, John. 1974. The glass house. Cambridge, Mass: MIT Press.
32. Pati, Ranjan; Abelar, Michael (27 May 2015). "The Application and Optimization of Metal Reflectors to Vertical Greenhouses to Increase Plant Growth and Health". Journal of Agricultural Engineering and Biotechnology: 63–71. doi:10.18005/JAEB0302003.
33. "Glossary for Vertical Farming". Retrieved 2016-01-06.
34. Vertical farming (1915). Wilmington, Del.: E. I. duPont de Nemours Powder Co. Retrieved 2011-07-23.
35. Venkataraman, Bina (2008-07-15). "Country, the City Version: Farms in the Sky Gain New Interest". The New York Times. Retrieved 2011-01-05.
36. https://www.youtube.com/watch?v=1t-PUIXUPgw
37. https://www.youtube.com/watch?v=zWommg6JVfY
38. https://www.youtube.com/watch?v=r-Iymp-gdbo
39. https://www.youtube.com/watch?v=PrLTuGKcvPA
40. "The Rise of Urban Farming". Smithsonian Magazine. Retrieved 2011-01-05.
41. Douglas, James Sholto (1975). Hydroponics: The Bengal System(5th ed.). New Dehli: Oxford University Press. ISBN 9780195605662.
42. Nelson, Bryn (2007-12-12). "Could vertical farming be the future? Nelson, B. (2008).". MSNBC. Retrieved 2010-11-10.
43. George Monbiot (2010-08-16). "Monbiot, G. (16 August 2010). Greens living in ivory towers now want to farm them too. 'The". London: Guardian. Retrieved 2010-11-10.
44. https://store.ctbuh.org/PDF_Previews/Journal/CTBUHJournal_2009-2.pdf
45. Ruby, Andreas. Groundscapes pp. 87–93
46. Eaton, Ruth. Ideal Cities p. 239
47. link Archived August 11, 2011, at the Wayback Machine.
48. http://www.impactlab.net/wp-content/uploads/2008/10/editt-tower-by-tr-hamzah-and-yeang.jpg
49. http://ps-survival.com/PS/Hydroponics/Hydroponics_The_Bengal_System_1959.

pdf

50. Arnie Cooper (May 19, 2009). "Going up? Farming in High Rises Raises Hopes". Miller-mccune.com. Retrieved 2010-11-10.
51. Whyte, Murray (2008-07-27). "Is high rise farming in Toronto's future?". Toronto Star. Retrieved 2008-08-12.
52. "Sky Farm Proposed for Downtown Toronto". TreeHugger. Retrieved 2009-03-14.
53. Venkataraman, Bina (2008-07-15). "Country, the City Version: Farms in the Sky Gain New Interest". The New York Times.
54. Shute, Nancy (2007-05-20). "Farm of the Future? Someday food may grow in skyscrapers". U.S. News & World Report.
55. Feldman, Amy (2007-07-11). "Skyscraper Farms". Popular Science.
56. Despommier, Dickson (November 2009). "The Rise of Vertical Farms". Scientific American. New york: Scientific American Inc. 301 (5): 60–67. ISSN 0036-8733.
57. http://inhabitat.com/vertical-harvest-of-jackson-hole-will-grow-organic-produce-even-in-the-winter/
58. "Vertical Fresh Farms LLC, Buffalo, NY | StateLog". www.statelog.com. Retrieved 2015-11-05.
59. Clive Clifton says: (2009-08-24). "Vertical Farming: Too Far Outside the Box? |". E4capital.com. Retrieved 2010-11-10.
60. Michael Bomford says: (2010-12-02). "Energy and VerticalFarms". Retrieved 2011-03-08.
61. Evans, P. (July 22, 2009). Local food no green panacea: professor. CBC News http://www.cbc.ca/consumer/story/2009/07/22/consumer-local-food.html
62. https://www.youtube.com/watch?v=KZN3ucwePGk
63. "Does it really stack up?". The Economist. 2010-12-09.
64. "Starting a Commercial Greenhouse Business". Omafra.gov.on.ca. 2003-07-28. Retrieved 2010-11-10.
65. Pocket World in Figures, The Economist, 2011 ed. pg 64
66. Roach, J. (June 30, 2009). "High-Rise Farms: The Future of Food?". National Geographic News.
67. "George Monbiot – Towering Lunacy". Monbiot.com. Retrieved 2013-06-12.
68. The Economist (December 9, 2010). "Vertical farming: Does it really stack up?".
69. "Crops | Greenhouse | Greenhouse Energy Calculations | Manitoba Agriculture, Food and Rural Initiatives | Province of Manitoba". Gov.mb.ca. Retrieved 2010-11-10.
70. "About The Plant". The Plant.
71. "Archived copy" (PDF). Archived from the original (PDF) on 2011-07-19. Retrieved 2010-11-09.
72. Blom, T.J.; W.A. Straver; F.J. Ingratta; Shalin Khosla; Wayne Brown (December

2002). "Carbon Dioxide In Greenhouses". Retrieved 2010-10-10.

73. Light pollution handbook – Google Books. Books.google.ca. 2004. ISBN 978-1-4020-2665-2. Retrieved 2010-11-10.
74. "Treating and Recycling Irrigation Runoff". Aggie-horticulture.tamu.edu. Retrieved 2013-06-12.
75. Despommier, D. (2008). "Vertical Farm Essay I". Vertical Farm. Retrieved 2009-06-26.
76. The Vertical Farm Project. 2009. "Agriculture for the 21st Century and Beyond". http://www.verticalfarm.com/
77. Frediani, K. L. 2010 Feeding time at the Zoo. The Horticulturalist. April 2010, pp.12–15 Institute of Horticulture.
78. Frediani, K. L. (2011), "High rise food", The Horticulturalist, October, pp.18–20.
79. Despommier, D. (2008). "Vertical Farm Essay I". Vertical Farm. Retrieved 2009-06-26.
80. "Vertical Farm Video". Discovery Channel. 2009-04-23. Retrieved 2009-06-26.
81. "Dwarf Wheat grown aboard the International Space Station". NASA. 9 February 2003. Retrieved 17 November 2009.
82. Michael Pollan (September 9, 2009). "Big Food vs. Big Insurance". The New York Times.
83. "A Farm on Every Floor", The New York Times, August 23, 2009
84. Vertical take off, Fresh Produce Journal, 28 January 2011
85. S.L. Davis (2001). "The least harm principle suggests that humans should eat beef, lamb, dairy, not a vegan diet". Proceedings of the Third Congress of the European Society for Agricultural and Food Ethics. pp. 449–450.
86. Navarro, L.M. & Pereira, H.M. (2012). "Rewilding Abandoned Landscapes in Europe.": 900–912.
87. "Food and Culture Spring 2011 course". ocw.mit.edu.
88. Case Study — Landfill Power Generation, H. Scott Matthews, Green Design Initiative, Carnegie Mellon University. "Archived copy" (PDF). Archived from the original (PDF) on 2008-12-03. Retrieved 2009-02-09. Retrieved 07.02.09
89. Folke Günther (2013-01-06). "The folkewall, greywater purification AND vertical growing". Holon.se. Retrieved 2013-06-12.
90. McConnell, Kathryn (2008-07-01). "Vertical Farms Grow Food by Growing Up, Not Out". Bureau of International Information Programs. United States Department of State. Retrieved 2008-08-12.
91. "Las Vegas to Build World's First 30 Story Vertical Farm". Nextenergynews.com. Retrieved 2013-06-12.
92. Despommier, Dickson (2010-06-15). "Dickson Despommier. November 2009. "Growing Skyscrapers: The Rise of Vertical Farms."". Scientificamerican.com. Retrieved 2010-11-10.

93. "Vertical plant production as a public exhibit at Paignton Zoo" (PDF). Proceedings of the 4th World Botanic Garden Congress, Botanic Gardens Conservation International. June 2010. Retrieved 2013-01-27.
94. "Green Zionist Alliance (GZA) – Bold Resolutions for 36th World Zionist Congress".
95. "World's first commercial vertical farm opens in Singapore". Io9.com. Retrieved 2013-06-12.
96. "Urban farming looking up in Singapore – CNN.com". Edition.cnn.com. Retrieved 2013-06-12.
97. "vertical farming infoday munich - agritecture.com". Retrieved 2015-06-01.
98. "Vertical Farming technologies trends - agritecture.com". Retrieved 2015-06-01.
99. Natural Landscaping: Designing With Native Plant Communities - John Diekelmann, Robert M. Schuster - Google Books. Books.google.co.uk. Retrieved 2013-04-10.
100. Landscaping Principles and Practices - Jack Ingels - Google Books. Books. google.co.uk. 2009-01-15. Retrieved 2013-04-10.
101. Landscaping - William Slack - Google Books. Books.google.co.uk. Retrieved 2013-04-10.
102. Taylor's Master Guide to Landscaping - Rita Buchanan - Google Books. Books. google.co.uk. Retrieved 2013-04-10.
103. John Smith. "Landscaping by landscape gardeners: Methods and Tactics". New Ways Landscaping Design. Retrieved 2016-06-14.
104. Sharon Cohoon and Jim McCausland. "How to Landscape Gravel - Page 2". Sunset.com. Archived from the original on March 19, 2012. Retrieved 2013-04-10.

CHAPTER 4

Hanging Garden Cultivation

Chapter Contents

- **Introduction**
- **Modern**
- **Hanging Gardens of Babylon**

INTRODUCTION

A hanging garden is a sustainable landscape architecture, an artistic garden or a small urban farm: attached to or built on a wall. They mainly found in areas where land is scarce or where the farmer is mobile or not permanent.

The most famous hanging gardens were the legendary Hanging Gardens of Babylon. They were considered in antiquity as one of the Seven Wonders of the World, and were located in present day Iraq.

MODERN

In contemporary use, hanging gardens are a green wall on a ground level facade, a balcony, a terrace, or part of a roof garden; of a home, or skyrise greenery with a residential, commercial, or government office building.

Products

Prefabricated modular hanging wall garden systems have been developed and are on the market internationally.

HANGING GARDENS OF BABYLON

The Hanging Gardens of Babylon, one of the Seven Wonders of the Ancient World, is the only one whose location has not been definitively established.

The Hanging Gardens were described as a remarkable feat of engineering: an ascending series of tiered gardens containing all manner of trees, shrubs, and vines. The gardens were said to have looked like a large green mountain constructed of mud bricks.

Traditionally they were said to have been built in the ancient city of Babylon, near present-day Hillah, Babil province, in Iraq. The Babylonian priest Berossus, writing in about 290 BC and quoted later by Josephus, attributed the gardens to the Neo-Babylonian king Nebuchadnezzar II, who ruled between 605 and 562 BC. There are no extant Babylonian texts which mention the gardens, and no definitive archaeological evidence has been found in Babylon.

According to one legend, Nebuchadnezzar II built the Hanging Gardens for his Median wife, Queen Amytis, because she missed the green hills and valleys of her homeland. He also built a grand palace that came to be known as The Marvel of the Mankind.

Because no physical evidence for the Hanging Gardens has be found at Babylon it has been suggested that they were purely mythical, and the descriptions found in ancient Greek and Roman writers including Strabo, Diodorus Siculus and Quintus Curtius Rufus represent a romantic ideal of an eastern garden. If it did indeed exist, it was destroyed sometime after the first century AD.

Stephanie Dalley suggests that the original garden may have been a well-documented one that the Assyrian king Sennacherib (704–681 BC) built in his capital city of Nineveh on the River Tigris, near the modern city of Mosul.

This hand-coloured engraving, probably made in the 19th century after the first excavations in the Assyrian capitals, depicts the fabled Hanging Gardens, with the Tower of Babel in the background

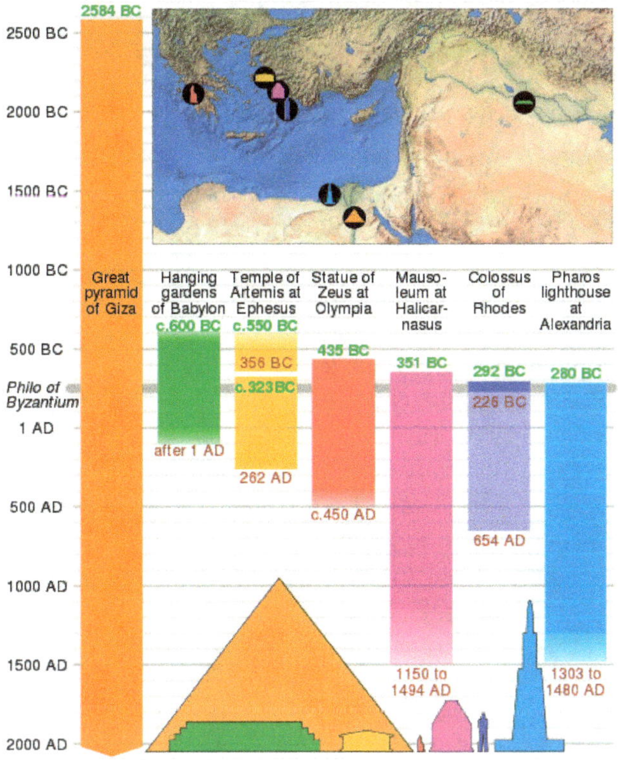

Timeline and map of the *Hanging Gardens of Babylon* and the other Wonders of the Ancient World

Ancient Texts

There are five principal writers whose descriptions of Babylon are extant in some form today. These writers concern themselves with the size of the Hanging Gardens, their overall design and means of irrigation, and why they were built.

Josephus (c.37–100 AD) quotes a description of the gardens by Berossus, a Babylonian priest of Marduk writing circa 290 BC. Berossus described the reign of Nebuchadnezzar II, and is the only source to credit that king with the construction of the Hanging Gardens.

In this palace he erected very high walks, supported by stone pillars; and by planting what was called a pensile paradise, and replenishing it with all sorts of trees, he rendered the prospect an exact resemblance of a mountainous country. This he did to gratify his queen, because she had been brought up in Media, and was fond of a mountainous situation.

For his description of the gardens, Diodorus Siculus (active c.60–30 BC) seems to have consulted the 4th century BC texts of both Cleitarchus (a historian of Alexander the Great) and Ctesias of Cnidus. Diodorus ascribes the construction to a Syrian king.

The park extended four plethra on each side, and since the approach to the garden sloped like a hillside and the several parts of the structure rose from one another tier on tier, the appearance of the whole resembled that of a theatre. When the ascending terraces had been built, there had been constructed beneath them galleries which carried the entire weight of the planted garden and rose little by little one above the other along the approach; and the uppermost gallery, which was fifty cubits high, bore the highest surface of the park, which was made level with the circuit wall of the battlements of the city. Furthermore, the walls, which had been constructed at great expense, were twenty-two feet thick, while the passageway between each two walls was ten feet wide. The roof above these beams had first a layer of reeds laid in great quantities of bitumen, over this two courses of baked brick bonded by cement, and as a third layer of covering of lead, to the end that the moisture from the soil might not penetrate beneath. On all this again earth had been piled to a depth sufficient for the roots of the largest trees; and the ground, when levelled off, was thickly planted with trees of every kind that, by their great size or other charm, could give pleasure to the beholder. And since the galleries, each projecting beyond another, all received the light, they contained many royal lodgings of every description; and there was one gallery which contained openings leading from the topmost surface and machines for supplying the gardens with water, the machines raising the water in great abundance from the river, although no one outside could see it being done. Now this park, as I have said, was a later construction.

The next account is from Quintus Curtius Rufus (active 1st century AD), who probably drew on the same sources as Diodorus.

The Babylonians also have a citadel twenty stades in circumference. The foundations of its turrets are sunk thirty feet into the ground and the fortifications rise eighty feet above it at the highest point. On its summit are the hanging gardens, a wonder celebrated by the fables of the Greeks. They are as high as the top of the walls and owe their charm to the shade of many tall trees. The columns supporting the whole edifice are built of rock, and on top of them is a flat surface of squared stones strong enough to bear the deep layer of earth placed upon it and the water used for irrigating it. So stout are the trees the structure supports that their trunks are eight cubits thick and their height as much as fifty feet; they bear fruit as abundantly as if they were growing in their natural environment. And although time with its gradual decaying processes is as destructive to nature's works as to man's, even so this edifice survives undamaged, despite being subjected to the pressure of so many tree-roots and the strain of bearing the weight of such a huge forest. It has a substructure of walls twenty feet thick at eleven foot intervals, so that from a distance one has the impression of woods overhanging their native mountains. Tradition has it that it is the work of a Syrian king who ruled from Babylon. He built it out of love for his wife who missed the woods and forests in this flat country and persuaded her husband to imitate nature's beauty with a structure of this kind.

Strabo (c.64 BC – 21 AD) described the Hanging Gardens as follows, in a passage thought to be based on the lost account of Onesicritus from the 4th century BC:

Babylon, too, lies in a plain; and the circuit of its wall is three hundred and eighty-five stadia. The thickness of its wall is thirty-two feet; the height thereof between the towers is fifty cubits; that of the towers is sixty cubits; and the passage on top of the wall is such that four-horse chariots can easily pass one another; and it is on this account that this and the hanging garden are called one of the Seven Wonders of the World. The garden is quadrangular in shape, and each side is four plethra in length. It consists of arched vaults, which are situated, one after another, on checkered, cube-like foundations. The checkered foundations, which are hollowed out, are covered so deep with earth that they admit of the largest of trees, having been constructed of baked brick and asphalt – the foundations themselves and the vaults and the arches. The ascent to the uppermost terrace-roofs is made by a stairway; and alongside these stairs there were screws, through which the water was continually conducted up into the garden from the Euphrates by those appointed for this purpose, for the river, a stadium in width, flows through the middle of the city; and the garden is on the bank of the river.

Philo of Byzantium (4th–5th century AD) is the last of the classical sources. The description of the gardens in his *A Handbook to the Seven Wonders of the World* is thought to be independent of earlier Greek sources. The method of raising water by screw matches that described by Strabo.

The so-called Hanging Gardens have plants above ground, and are cultivated in the air, with the roots of the trees above the (normal) tilled earth, forming a roof. Four stone columns are set beneath, so that the entire space through the carved pillars is beneath the (artificial) ground. Palm trees lie in place on top of the pillars, alongside each other as (cross-) beams, leaving very little space in between. This timber does not rot, unlike others; when it is soaked and put under pressure it swells up and nourishes the growth from roots, since it incorporates into its own interstices what is planted with it from outside. Much deep soil is piled on, and then broad-leaved and especially garden trees of many varieties are planted, and all kind of flowering plants, everything, in short, that is most joyous and pleasurable to the onlooker. The place is cultivated as if it were (normal) tilled earth, and the growth of new shoots has to be pruned almost as much as on normal land. This (artificial) arable land is above the heads of those who stroll along through the pillars. When the uppermost surface is walked on, the earth on the roofing stays firm and undisturbed just like a (normal) place with deep soil. Aqueducts contain water running from higher places; partly they allow the flow to run straight downhill, and partly they force it up, running backwards, by means of a screw; through mechanical pressure they force it round and round the spirals of the machines. Being discharged into close-packed, large cisterns, altogether they irrigate the whole garden, inebriating the roots of the plants to their depths, and maintaining the wet arable land, so that it is just like an evergreen meadow, and the leaves of the trees, on the tender new growth, feed upon dew and have a wind-swept appearance. For the roots, suffering no thirst, sprout anew, benefitting from the moisture of the water that runs past, flowing at random, interweaving along the lower ground to the collecting point, and reliably protects the growing of trees that have become established. Exuberant and fit for a king is the ingenuity, and most of all, forced, because the cultivator's hard work is hanging over the heads of the spectators.

Scholarship and Controversy

This copy of a bas relief from the North Palace of Ashurbanipal (669–631 BC) at Nineveh shows a luxurious garden watered by an aqueduct.

There is some controversy as to whether the Hanging Gardens were an actual construction or a poetic creation, owing to the lack of documentation in contemporaneous Babylonian sources. There is also no mention of Nebuchadnezzar's wife Amyitis (or any other wives), although a political marriage to a Median or Persian would not have been unusual. Many records exist of Nebuchadnezzar's works, yet his long and complete inscriptions do not mention any garden.

Herodotus, who describes Babylon in his *Histories*, does not mention the Hanging Gardens.

To date, no archaeological evidence has been found at Babylon for the Hanging Gardens. It is possible that evidence exists beneath the Euphrates, which cannot be excavated safely at present. The river flowed east of its current position during the time of Nebuchadnezzar II, and little is known about the western portion of Babylon.Rollinger has suggested that Berossus attributed the Gardens to Nebuchadnezzar for political reasons, and that he had adopted the legend from elsewhere.

A recent theory proposes that the Hanging Gardens of Babylon were actually constructed by the Assyrian king Sennacherib (reigned 704 – 681 BC) for his palace at Nineveh. Stephanie Dalley posits that during the intervening centuries the two sites became confused, and the extensive gardens at Sennacherib's palace were attributed to Nebuchadnezzar II's Babylon. Recently discovered evidence includes excavation of a vast system of aqueducts inscribed to Sennacherib, which Dalley proposes were part of a 80-kilometre (50 mi) series of canals, dams, aqueducts, used to carry water to Nineveh with water-raising screws used to raise it to the upper levels of the gardens.

Dalley bases her arguments on recent developments in the decipherment of contemporary Akkadian inscriptions. Her main points are:

- The name Babylon, meaning Gate of the Gods was applied to several Mesopotamian cities. Sennacherib renamed the city gates of Nineveh after gods, which suggests that he wished his city to be considered a Babylon.
- Only Josephus names Nebuchadnezzar as the king who built the gardens, but although Nebuchadnezzar left many inscriptions none mentions any garden or engineering works. Diodorus Siculus and Quintus Curtius Rufus specify a Syrian king.
- By contrast Sennacherib left written descriptions and there is archaeological evidence of his water engineering. His grandson Assurbanipal pictured the mature garden on a sculptured wall panel in his palace.
- Sennacherib called his new palace and garden a wonder for all peoples. He describes the making and operation of screws to raise water in his garden.
- The descriptions of the classical authors fit closely to these contemporary records. Before the Battle of Gaugamela in 331 BC Alexander the Great camped for four days near the aqueduct at Jerwan. The historians who travelled with him would have had ample time to investigate the enormous works around them, recording them in Greek. These first-hand accounts do not survive into our times, but were quoted by later Greek writers.

Hanging Garden at Nineveh

King Sennacherib's Hanging Garden was considered a World Wonder not just for its beauty – a year-round oasis of lush green in a dusty summer landscape – but also for the marvellous feats of water engineering that maintained the garden.

There was a tradition of Assyrian royal garden building. King Ashurnasirpal II (883–859 BC) describes what he had done:

I dug out a canal from the (river) Upper Zab, cutting through a mountain peak, and called it the Abundance Canal. I watered the meadows of the Tigris and planted orchards with all kinds of fruit trees in the vicinity. I planted seeds and plants that I had found in the countries through which I had marched and in the highlands which I had crossed: pines of different kinds, cypresses and junipers of different kinds, almonds, dates, ebony, rosewood, olive, oak, tamarisk, walnut, terebinth and ash, fir, pomegranate, pear, quince, fig, grapevine.... The canal water gushes from above into the garden; fragrance pervades the walkways, streams of water as numerous as the stars of heaven flow in the pleasure garden.... Like a squirrel I pick fruit in the garden of delights...

Sennacherib is the only Mesopotamian king who has left a record of his love for his wife – a key part of the romantic classical story:

And for Tashmetu-sharrat the palace woman, my beloved wife, whose features the Mistress of the Gods has made perfect above all other women, I had a palace of loveliness, delight and joy built...

Sennacherib's palace was comparable in size to Windsor Castle in England. He specifically mentions the massive limestone blocks that reinforce the flood defences. Parts of the palace were excavated by Austin Henry Layard in the mid-19th century. His citadel plan shows contours which would be consistent with Sennacherib's garden, but its position has not been confirmed. The area has been used as a military base in recent times, making it difficult to investigate further.

A sculptured wall panel of Assurbanipal shows the garden in its maturity. There is one original panel and the drawing of another in the British Museum, although neither is on public display. Several features mentioned by the classical authors are discernible on these contemporary images.

The irrigation of such a garden demanded an upgraded water supply to the city of Nineveh. The canals stretched over 50 km into the mountains. Sennacherib was proud of the technologies he had employed, and describes them in some detail on his inscriptions. For example:

At the headwater of Bavian (Khinnis) his inscription mentions automatic sluice gates but does not say how they worked:

The sluice gate of that canal opens without a spade or a shovel and lets the waters of abundance flow.: its sluice gate is not opened by the labour of men's hands, but by the will of the gods.

An enormous aqueduct crossing the valley at Jerwan was constructed of over 2 million dressed stones. It used stone arches and waterproof cement. On it is written:

Sennacherib king of the world king of Assyria. Over a great distance I had a watercourse directed to the environs of Nineveh, joining together the waters.... Over steep-sided valleys I spanned an aqueduct of white limestone blocks, I made those waters flow over it.

He claims to be the first to deploy a new casting technique in place of the lost-wax process for his monumental (30 tonne) bronze castings, and describes the making of his water screws (though once again he does not say exactly how they were driven):

Whereas in former times the kings my forefathers had created bronze statues imitating real-life forms to put on display inside temples, but in their method of work they had exhausted all the craftsmen, for lack of skill and failure to understand the principles they needed so much oil, wax and tallow for the work that they caused a shortage in their own countries – I Sennacherib, leader of all princes, knowledgeable in all kinds of work, took much advice and deep thought over doing that kind of work.... I created clay moulds as if by divine intelligence for cylinders and screws... In order to draw up water all day long, I had ropes, bronze wires and bronze chains made. And instead of a shaduf I set up the cylinders and screws of copper over cisterns....I raised the height of the surroundings of the palace, to be a Wonder for all Peoples... A high garden imitating the Amanus mountains I laid out next to it, with all kinds of aromatic plants, orchard fruit trees, trees that enrich not only mountain country but also Chaldea (Babylonia), as well as trees that bear wool, planted within it.

Sennacherib could bring the water into his garden at a high level because it was sourced from further up the mountains. He then raised the water even higher by deploying his new water screws. This meant he could build a garden that towered into the sky with large trees on the top of the terraces – a stunning artistic effect that surpassed those of his predecessors and which justifies his own claim to have built a Wonder for all Peoples.

REFERENCES

1. Finkel (1988) p. 41.
2. Finkel (1988) p. 58.
3. Finkel, Irving; Seymour, Michael (2008). Babylon: City of Wonders. London: British Museum Press. p. 52. ISBN 0-7141-1171-6.
4. Finkel 2008
5. "The Hanging Gardens of Babylon". Retrieved 5 February 2014.
6. "The Hanging Gardens of Babylon". Retrieved 5 February 2014.
7. Dalley, Stephanie (2013). The Mystery of the Hanging Garden of Babylon: an elusive World Wonder traced. Oxford University Press. ISBN 978-0-19-966226-5.
8. Finkel (2008) p. 108.
9. Joseph. contr. Appion. lib. 1. c. 19.—Syncel. Chron. 220.—Euseb. Præp. Evan. lib. 9.
10. Diodorus Siculus II.10-1-10
11. History of Alexander V.1.35-5
12. Strabo, Geography XVI.1.5, translation adapted from H.L. Jones, Loeb Classical Library edn (1961).
13. This author is now thought to not be Philo the Engineer of Byzantium, but perhaps Philo the Paradoxographer of Byzantium; Stephanie Dalley, "More about the Hanging Gardens," in Of Pots and Pans: Papers on the Archaeology and History of Mesopotamia and Syria as presented to David Oates on his 75th Birthday, Edited by L. al-Gailani-Werr, J.E. Curtis, H. Martin, A. McMahon, J. Oates and J.E. Reade, (London), pp. 67–73 ISBN 1-897750-62-5.
14. Dalley (2013), p. 40. Dalley bases her translation on Brodersen (1992) who uses an early Greek text. A previous translation by David Oates, based on a Latin text, is found in Finkel (1988) pp. 45–46.
15. Finkel (2008) p. 109.
16. Dalley (2013)
17. Priestley, Jessica (2014). Herodotus and Hellenistic culture: Literary Studies in the Reception of the Histories. Oxford: Oxford University Press. p. 91.
18. Joan Oates, Babylon, Revised Edition, Thames and Hudson, London (1986) p. 144 ISBN 0500273847.
19. Rollinger, Robert "Berossos and the Monuments", ed. J Haubold et al, The World of Berossos, Wiesbaden (2013), p151
20. Stephanie Dalley (1993). "Ancient Mesopotamian Gardens and the Identification of the Hanging Gardens of Babylon Resolved". Garden History. 21: 7. JSTOR 1587050.
21. Alberge, Dalya (5 May 2013). "Babylon's hanging garden: ancient scripts give clue to missing wonder". The Guardian. Retrieved 6 May 2013.

22. Dalley, Stephanie (2013) The Mystery of the Hanging Garden of Babylon: an elusive World Wonder traced, Oxford University Press ISBN 978-0-19-966226-5.
23. AR George, Babylonian Topographical Texts, (1992)
24. see for example Cuneiform Texts in the British Museum, Vol 19, page 25, line 25
25. Pongratz-Leisten, Ina Sulmi Erub (1994),
26. See Dalley (2013) ch 1 for a summary.
27. Especially: the Iraq Museum prism dated 694 BC published by A Heidel, The Octagonal Sennacherib Prism in the Iraq Museum, Sumer 9 (1953); and the British Museum prism BM103000 of the same date
28. T Jacobsen and S Lloyd, Sennacherib's Aqueduct at Jerwan (1935); Reade, Studies in Assyrian Geography, Revue d'Assyriologie 72 (1978); Channel 4 TV programme Secret History: Finding Babylon's Hanging Garden, 24th November 2013
29. AH Layard, Discoveries in the Ruins of Nineveh and Babylon, (1853)
30. Dalley (2013), pp. 62–63
31. R Lane Fox, Alexander the Great (1973)
32. Dalley, Stephanie. The Mystery of the Hanging Garden of Babylon; an elusive World Wonder traced, Oxford University Press (2013). ISBN 978-0-19-966226-5. The quotations in this section are the translations of the author and are reproduced with the permission of OUP.
33. BM124939
34. Original Drawing IV 77
35. Layard (1853)
36. Jacobsen (1935)

CHAPTER 5

Integrated Farming

Chapter Contents

- **Introduction**
- **Definition**
- **Classification**
- **History**
- **Objectives**
- **Prevalence**
- **Criticism**
- **Holistic Management (Agriculture)**
- **Renewable Energy**

Integrated Farming is a whole farm management system, which enables the farmers to identify opportunities and threats and act accordingly, and, at the same time, consider consumer interests in their business.

Integrated Farming is not based on a set of fixed parameters but on informed management processes.

INTRODUCTION

Integrated Farming (IF) or integrated production is a whole farm management system which aims to deliver more sustainable agriculture. It is a dynamic approach which can be applied to any farming system around the world. It involves attention to detail and continuous improvement in all areas of a farming business through informed management processes. Integrated Farming combines the best of modern tools and technologies with traditional practices according to a given site and situation. In simple words, it means using many ways of cultivation in a small space or land.

DEFINITION

The holistic approach: Integrated Farming looks at and relates to the whole farm.

The International Organisation of Biological Control (IOBC) describes Integrated Farming as a farming system where high quality food, feed, fibre and renewable energy are produced by using resources such as soil, water, air and nature as well as regulating factors to farm sustainably and with as little polluting inputs as possible.

Particular emphasis is placed on a holistic management approach looking at the whole farm as cross-linked unit, on the fundamental role and function of agro-ecosystems, on nutrient cycles which are balanced and adapted to the demand of the crops, and on health and welfare of all livestock on the farm. Preserving and enhancing soil fertility, maintaining and improving a diverse environment and the adherence to ethical and social criteria are indispensable basic elements. Crop protection takes into account all biological, technical and chemical methods which then are balanced carefully and with the objective to protect the environment, to maintain profitability of the business and fulfil social requirements.

EISA European Initiative for Sustainable Development in Agriculture e. V. have an Integrated Farming Framework which provides additional explanations on key aspects of Integrated Farming. These include: Organisation & Planning, Human & Social Capital, Energy Efficiency, Water Use & Protection, Climate Change & Air Quality, Soil Management, Crop Nutrition, Crop Health & Protection, Animal Husbandry, Health & Welfare, Landscape & Nature Conservation and Waste Management Pollution Control.

LEAF (Linking Environment and Farming) in the UK promotes a comparable model and defines Integrated Farm Management (IFM) as whole farm business approach that delivers sustainable farming.

CLASSIFICATION

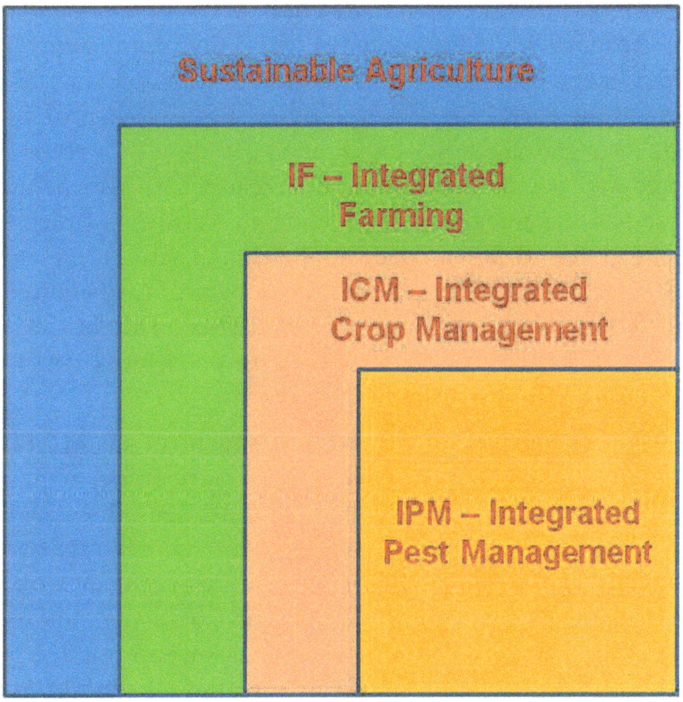

Integrated Farming in the context of sustainable agriculture.

The Food and Agriculture Organization of the United Nations FAO promotes Integrated Pest Management (IPM) as the preferred approach to crop protection and regards it as a pillar of both sustainable intensification of crop production and pesticide risk reduction". IPM thus is one indispensable element of Integrated Crop Management which in turn is one essential part of the holistic Integrated Farming approach towards sustainable agriculture.

KELLER, 1985 (quoted in Lütke Entrup et al., 1998 1) highlights that Integrated Crop Management is not to be understood as compromise between different agricultural production systems. It rather must be understood as production system with a targeted, dynamic and continuous use and development of experiences which were made in the so-called conventional farming. In addition to natural scientific findings, impulses from organic farming are also taken up.

HISTORY

Integrated Pest Management can be seen as starting point for a holistic approach to agricultural production. Following the excessive use of crop protection chemicals, first steps in IPM were taken in fruit production at the end of the 1950s. The concept was then further developed globally in all major crops. On the basis of results of the system-oriented IPM approach, models for Integrated Crop Management were developed. Initially, animal husbandry was not seen as part of such integrated approaches (Lütke Entrup et al., 1998 1).

In the years to follow, various national and regional initiatives and projects were formed. These include LEAF (Linking Environment And Farming) in the UK, FNL (Fördergemeinschaft Nachhaltige Landwirtschaft e.V.) in Germany, FARRE (Forum des Agriculteurs Responsables Respectueux de l'Environnement) in France, FILL (Fördergemeinschaft Integrierte Landbewirtschaftung Luxemburg) or OiB (Odling i Balans) in Sweden. However, there are few if any figures on the uptake of Integrated Farming in the major crops throughout Europe for example, leading to a recommendation by the European Economic and Social Committee in February 2014, that the EU should carry out an in-depth analysis of integrated production in Europe in order to obtain insights into the current situation and potential developments. There is evidence, however, that between 60 and 80% of pome, stone and soft fruits were grown, controlled and marketed according to "Integrated Production Guidelines" in 1999 already in Germany for example. In the UK, 22% of fresh fruit and vegetables are grown to Integrated Farm Management standards as recognised by the LEAF Marque.

Animal husbandry and Integrated Crop Management (ICM) often are just two branches of one agricultural enterprise. In modern agriculture, animal husbandry and crop production must be understood as interlinked sectors which cannot be looked at in isolation, as the context of agricultural systems leads to tight interdependencies. Uncoupling animal husbandry from arable production (too high stocking rates) is therefore not considered in accordance with the principles and objectives of Integrated Farming (Lütke Entrup et al., 1998 1). Accordingly, holistic concepts for Integrated Farming or Integrated Farm Management such as the EISA Integrated Farming Framework, and the concept of sustainable agriculture are increasingly developed, promoted and implemented at the global level.

Related to the 'sustainable intensification' of agriculture, an objective which in part is discussed controversially, efficiency of resource use becomes increasingly important today. Environmental impacts of agricultural production depend on the efficiency achieved when using natural resources and all other means of production. The input per kg of output, the output per kg of input, and the output achieved per hectare of land – a limited resource in the light of world population growth – are decisive figures for evaluating the efficiency and the environmental impact of agricultural systems. Efficiency parameters therefore offer important evidence how efficiency and environmental impacts of agriculture can be judged and where improvements can or must be made.

Against this background, documentation as well certification schemes and farm audits such as LEAF Marque in the UK and 33 other countries throughout the world become more and more important tools to evaluate – and further improve – agricultural practices.

Even though being by far more product- or sector-oriented, SAI Platform principles and practices and GlobalGap for example pursue similar approaches.

OBJECTIVES

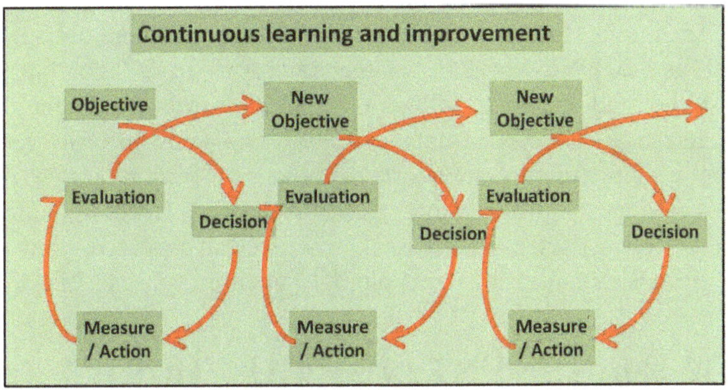

Continuous learning process in Integrated Farming.

Integrated Farming is based on attention to detail, continuous improvement and managing all resources available.

Being bound to sustainable development, the underlying three dimensions "economic development", "social development" and "environmental protection" are thoroughly considered in the practical implementation of Integrated Farming. However, the need for profitability is a decisive prerequisite: To be sustainable, the system must be profitable, as profits generate the possibility to support all activities outlined in the (EISA Integrated Farming) IF Framework.

As a management and planning approach, Integrated Farming includes regular benchmarking of targets set against results achieved. The concept of the EISA Integrated Farming Framework for example has a clear focus on farmers' awareness of their own performance. By regularly benchmarking their performance, farmers become aware of achievements as well as deficiencies, and by paying attention to detail they can continuously work on improving the whole farming enterprise and their economic performance at the same time: According to findings in UK, reducing fertiliser and chemical inputs to amounts according to the demand of the crops allowed for cost savings in the range of £2,500 - £10,000 per year and per farm.

PREVALENCE

Following first developments in the 1950s, various approaches to Integrated Pest Management, Integrated Crop Management, Integrated Production and Integrated Farming were developed worldwide (in Germany, Switzerland, US, Australia, and India, for example). As the implementation of the general concept of Integrated Farming and its individual components always should be handled according to the given site and situation instead of following strict rules and recipes, the concept is virtually applicable —— and being used to various degrees —— all over the world.

CRITICISM

It should be mentioned, however, that there are also critical voices from environmental organisations for example. That is in part due to the fact that there are European Organic Regulations such as (EC) No 834/2007 or the new draft from 2014 but no comparable regulations for Integrated Farming. Whereas organic farming and the "Bio-Siegel" in Germany for example are legally protected, EU Commission has not yet considered to start working on a comparable framework or blueprint for Integrated Farming. When products are marketed as "Controlled Integrated Produce", according control mechanisms and quality-labels are not based on national or European directives but are established and handled by private organisations and quality schemes such as LEAF Marque.

HOLISTIC MANAGEMENT (AGRICULTURE)

Holistic management in agriculture is a systems thinking approach to managing resources that was originally developed by Allan Savory for reversing desertification. In 2010 the Africa Centre for Holistic Management in Zimbabwe, Operation Hope (a proof of concept" project using holistic management) was named the winner of the 2010 Buckminster Fuller Challenge for "recognizing initiatives which take a comprehensive, anticipatory, design approach to radically advance human well being and the health of our planet's ecosystems."

Beginnings

The idea of holistic planned grazing began in the 1960s when Allan Savory, a wildlife biologist in his native Southern Rhodesia, set out to understand desertification. This can be seen in the context of the larger environmental movement. Heavily influenced by the work of André Voisin and the ineffectiveness of mainstream rangeland science of the time, Savory concluded that the spread of deserts, the loss of wildlife, and the human impoverishment that always resulted were related to the reduction of the natural herds of large grazers and even more, the change in behavior of those few remaining herds. Livestock could be substituted to provide important ecosystem services like nutrient cycling when mimicking those uniquely coevolved grasses and grazers. But managers had found that while rotational grazing systems can work for diverse management purposes, scientific experiments had demonstrated that they do not necessarily work for specific ecological purposes. An adaptive management plan was needed for the integration of the experiential with the experimental, as well as the social with the biophysical, to provide a more comprehensive framework for the management of rangeland systems. None of these sources of knowledge could be understood except in the context of the whole. Holistic management was developed to meet that need.

Development

In many regions, pastoralism and communal land use are blamed for environmental degradation caused by overgrazing. After years of research and experience, Savory came to understand this assertion was often wrong, and that sometimes removing animals actually made it worse. This concept is a variation of the trophic cascade, where humans are seen as the top level predator and the cascade follows from there.

I have been particularly fascinated, for example, by the work of a remarkable man called Allan Savory, in Zimbabwe and other semi-arid areas, who has argued for years against the prevailing expert view that is the simple numbers of cattle that drive overgrazing and cause fertile land to become desert. On the contrary, as he has since shown so graphically, the land needs the presence of feeding animals and their droppings for the cycle to be complete, so that soils and grassland areas stay productive. Such that, if you take grazers off the land and lock them away in vast feedlots, the land dies. - His Royal Highness The Prince of Wales (Prince Charles) from a speech to the IUCN World Conservation Congress

Savory developed a flexible management system designed to improve grazing systems. Holistic planned grazing is one of a number of newer grazing management systems that more closely simulate the behavior of natural herds of wildlife and have been shown to improve riparian habitats and water quality over systems that often led to land degradation, and be an effective tool to improve range condition for both livestock and wildlife. Holistic planned grazing is similar to rotational grazing but differs in that it more explicitly recognizes and provides a framework for adapting to the four basic ecosystem processes: the water cycle, the mineral cycle including the carbon cycle, energy flow, and community dynamics (the relationship between organisms in an ecosystem) as equal in importance to livestock production and social welfare. Thus the holistic context in the planning stage leads to different decisions in dealing with that complexity. Holistic management has been likened to a permaculture approach to rangeland management".

Uses

While originally developed as a tool for range land use and restoring desertified land, the holistic management system can be applied to other areas with multiple complex socioeconomic and environmental factors. One such example is Integrated Water Resources Management (IWRM), which promotes sector integration in development and management of water resources to ensure that water is allocated between different users in a fair way, maximizing economic and social welfare without compromising the sustainability of vital ecosystems. In essence, coordinated holistic water management takes into consideration all water users in nature and society. Another example is mine reclamation. A fourth use of Holistic management is in certain forms of no till crop production, intercropping, and permaculture. Holistic management has been acknowledged by The United States Department of Agriculture Natural Resources Conservation Service (USDA-NRCS).

Framework

The holistic management decision-making framework uses six key steps to guide the management of resources:

- Define in its entirety what you are managing. No area should be treated as a single-product system. By defining the whole, people are better able to manage. This includes identifying the available resources, including money, that the manager has at his disposal.
- Define what you want now and for the future. Set the objectives, goals and actions needed to produce the quality of life sought, and what the life-nurturing environment must be like to sustain that quality of life far into the future.
- Watch for the earliest indicators of ecosystem health. Identify the ecosystem services that have deep impacts for people in both urban and rural environments, and find a way to easily monitor them. One of the best examples of an early indicator of a poorly functioning environment is patches of bare ground. An indicator of a better functioning environment is newly sprouting diversity of plants and a return or increase of wildlife.
- Don't limit the management tools you use. The eight tools for managing natural resources are money/labor, human creativity, grazing, animal impact, fire, rest, living organisms and science/technology. To be successful you need to use all these tools to the best of your ability.
- Test your decisions with questions that are designed to help ensure all your decisions are socially, environmentally and financially sound for both the short and long term.
- Monitor proactively, before your managed system becomes more imbalanced. This way the manager can take adaptive corrective action quickly, before the ecosystem services are lost. Always assume your plan is less than perfect and use a feedback loop that includes monitoring for the earliest signs of failure, adjusting and re-planning as needed. In other words use a "canary in a coal mine" adaptive approach.

Four Principles

Holistic management planned grazing has four key principles that take advantage of the symbiotic relationship between large herds of grazing animals, their predators and the grasslands that support them:

- Nature functions as a holistic community with a mutualistic relationship between people, animals and the land. If you remove or change the behavior of any keystone species like the large grazing herds, you have an unexpected and wide ranging negative impact on other areas of the environment.
- It is absolutely crucial that any agricultural planning system must be flexible enough to adapt to nature's complexity, since all environments are different and have constantly changing local conditions.

- Animal husbandry using domestic species can be used as a substitute for lost keystone species. Thus when managed properly in a way that mimics nature, agriculture can heal the land and even benefit wildlife, while at the same time benefiting people.
- Time and timing is the most important factor when planning land use. Not only is it crucial to understand how long to use the land for agriculture and how long to rest, it is equally important to understand exactly when and where the land is ready for that use and rest.

Criticism

One limitation of any land management system is that economically and politically powerful users can easily quantify and argue their needs. It is harder to define the economic value of ecosystem services and, therefore, the ecosystems and people most dependent on them for their subsistence become voiceless and often neglected users. In theory Holistic Management framework addresses this issue, but it is not always seen in the field. Another common criticism of holistic planned grazing is that while farmers and ranchers around the world believe that it works for them and they have even received awards, the majority of range scientists have not been able to experimentally confirm that intensive grazing systems similar to those at the center of holistic management show a benefit, and claim that managers' reports of success are anecdotal. There is however a growing body of research work which is reversing the claim that the only evidence for holistic management is anecdotal.

RENEWABLE ENERGY

Wind, solar, and hydroelectricity are three emerging renewable sources of energy

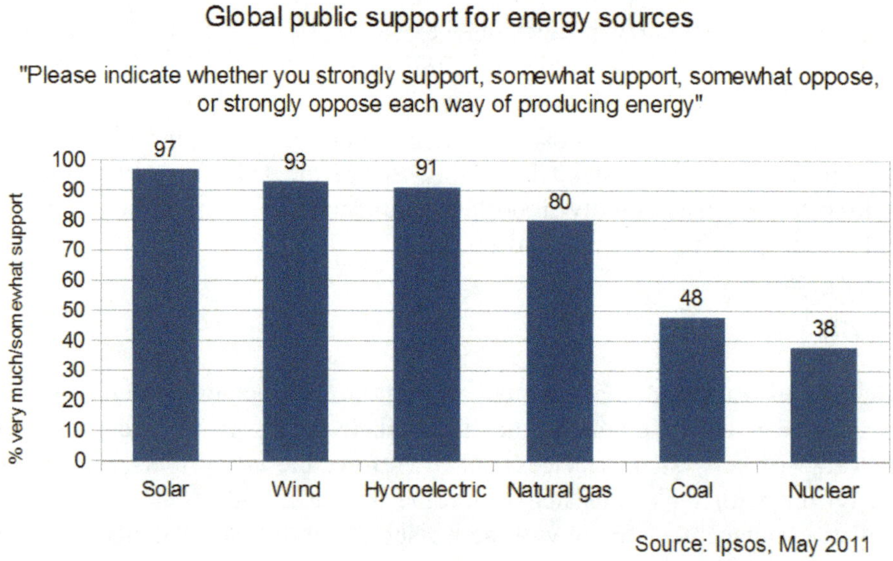

Global public support for different energy sources (2011) based on a poll by Ipsos Global @ dvisor

Renewable energy is energy that is collected from renewable resources, which are naturally replenished on a human timescale, such as sunlight, wind, rain, tides, waves, and geothermal heat. Renewable energy often provides energy in four important areas: electricity generation, air and water heating/cooling, transportation, and rural (off-grid) energy services.

Based on REN21's 2016 report, renewables contributed 19.2% to humans' global energy consumption and 23.7% to their generation of electricity in 2014 and 2015, respectively. This energy consumption is divided as 8.9% coming from traditional biomass, 4.2% as heat energy (modern biomass, geothermal and solar heat), 3.9% hydro electricity and 2.2% is electricity from wind, solar, geothermal, and biomass. Worldwide investments in renewable technologies amounted to more than US$286 billion in 2015, with countries like China and the United States heavily investing in wind, hydro, solar and biofuels. Globally, there are an estimated 7.7 million jobs associated with the renewable energy industries, with solar photovoltaics being the largest renewable employer. As of 2015 worldwide, more than half of all new electricity capacity installed was renewable.

Renewable energy resources exist over wide geographical areas, in contrast to other energy sources, which are concentrated in a limited number of countries. Rapid deployment of renewable energy and energy efficiency is resulting in significant energy security, climate change mitigation, and economic benefits. The results of a recent review of the literature concluded that as greenhouse gas (GHG) emitters begin to be held liable for damages resulting from GHG emissions resulting in climate change, a high value for liability mitigation would provide powerful incentives for deployment of renewable energy technologies. In international public opinion surveys there is strong support for promoting renewable sources such as solar power and wind power. At the national level, at least 30 nations around the world already have renewable energy contributing more than 20 percent of energy supply. National renewable energy markets are projected to continue to grow strongly in the coming decade and beyond. Some places and at least two countries, Iceland and Norway generate all their electricity using renewable energy already, and many other countries have the set a goal to reach 100% renewable energy in the future. For example, in Denmark the government decided to switch the total energy supply (electricity, mobility and heating/cooling) to 100% renewable energy by 2050.

While many renewable energy projects are large-scale, renewable technologies are also suited to rural and remote areas and developing countries, where energy is often crucial in human development. United Nations' Secretary-General Ban Ki-moon has said that renewable energy has the ability to lift the poorest nations to new levels of prosperity. As most of renewables provide electricity, renewable energy deployment is often applied in conjunction with further electrification, which has several benefits: Electricity can be converted to heat (where necessary generating higher temperatures than fossil fuels), can be converted into mechanical energy with high efficiency and is clean at the point of consumption. In addition to that electrification with renewable energy is much more efficient and therefore leads to a significant reduction in primary energy requirements, because most renewables don't have a steam cycle with high losses (fossil power plants usually have losses of 40 to 65%).

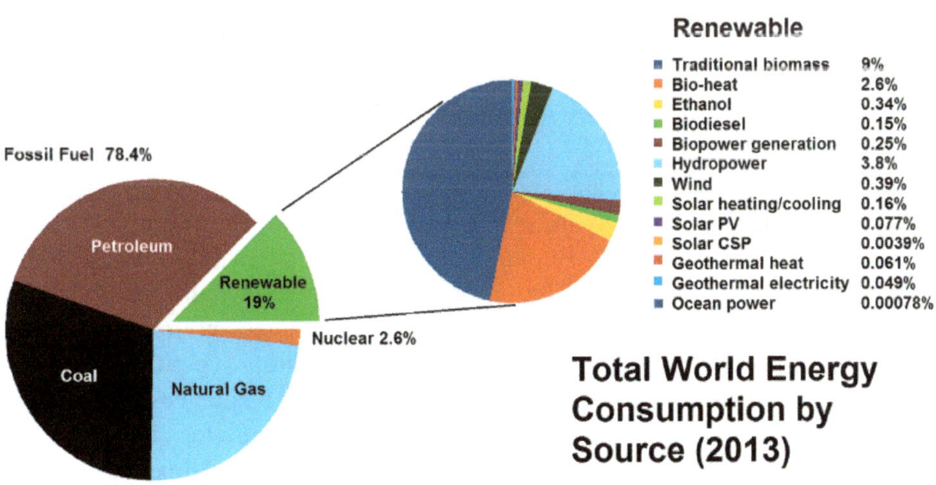

World energy consumption by source. Renewables accounted for 19% in 2012.

PlanetSolar, the world's largest solar-powered boat and the first ever solar electric vehicle to circumnavigate the globe (in 2012).

Renewable energy flows involve natural phenomena such as sunlight, wind, tides, plant growth, and geothermal heat, as the International Energy Agency explains:

Renewable energy is derived from natural processes that are replenished constantly. In its various forms, it derives directly from the sun, or from heat generated deep within the earth. Included in the definition is electricity and heat generated from solar, wind, ocean, hydropower, biomass, geothermal resources, and biofuels and hydrogen derived from renewable resources.

Renewable energy resources and significant opportunities for energy efficiency exist over wide geographical areas, in contrast to other energy sources, which are concentrated in a limited number of countries. Rapid deployment of renewable energy and energy efficiency, and technological diversification of energy sources, would result in significant energy security and economic benefits. It would also reduce environmental pollution such as air pollution caused by burning of fossil fuels and improve public health, reduce premature mortalities due to pollution and save associated health costs that amount to several hundred billion dollars annually only in the United States. Renewable energy sources, that derive their energy from the sun, either directly or indirectly, such as hydro and wind, are expected to be capable of supplying humanity energy for almost another 1 billion years, at which point the predicted increase in heat from the sun is expected to make the surface of the earth too hot for liquid water to exist.

Climate change and global warming concerns, coupled with high oil prices, peak oil, and increasing government support, are driving increasing renewable energy legislation, incentives and commercialization. New government spending, regulation and policies helped the industry weather the global financial crisis better than many other sectors. According to a 2011 projection by the International Energy Agency, solar power generators may produce most of the world's electricity within 50 years, reducing the emissions of greenhouse gases that harm the environment.

As of 2011, small solar PV systems provide electricity to a few million households, and micro-hydro configured into mini-grids serves many more. Over 44 million households use biogas made in household-scale digesters for lighting and/or cooking, and more than 166 million households rely on a new generation of more-efficient biomass cookstoves. United Nations' Secretary-General Ban Ki-moon has said that renewable energy has the ability to lift the poorest nations to new levels of prosperity. At the national level, at least 30 nations around the world already have renewable energy contributing more than 20% of energy supply. National renewable energy markets are projected to continue to grow strongly in the coming decade and beyond, and some 120 countries have various policy targets for longer-term shares of renewable energy, including a 20% target of all electricity generated for the European Union by 2020. Some countries have much higher long-term policy targets of up to 100% renewables. Outside Europe, a diverse group of 20 or more other countries target renewable energy shares in the 2020–2030 time frame that range from 10% to 50%.

Renewable energy often displaces conventional fuels in four areas: electricity generation, hot water/space heating, transportation, and rural (off-grid) energy services:

- Power generation

By 2040, renewable energy is projected to equal coal and natural gas electricity generation. Several jurisdictions, including Denmark, Germany, the state of South Australia and some US states have achieved high integration of variable renewables. For example, in 2015 wind power met 42% of electricity demand in Denmark, 23.2% in Portugal and 15.5% in Uruguay. Interconnectors enable countries to balance electricity systems by allowing the import and export of renewable energy. Innovative hybrid systems have emerged between countries and regions.

- Heating

Solar water heating makes an important contribution to renewable heat in many countries, most notably in China, which now has 70% of the global total (180 GWth). Most of these systems are installed on multi-family apartment buildings and meet a portion of the hot water needs of an estimated 50–60 million households in China. Worldwide, total installed solar water heating systems meet a portion of the water heating needs of over 70 million households. The use of biomass for heating continues to grow as well. In Sweden, national use of biomass energy has surpassed that of oil. Direct geothermal for heating is also growing rapidly. The newest addition to Heating is from Geothermal Heat Pumps which provide both heating and cooling, and also flatten the electric demand curve and are thus an increasing national priority (see also Renewable thermal energy).

- Transportation

Bioethanol is an alcohol made by fermentation, mostly from carbohydrates produced in sugar or starch crops such as corn, sugarcane, or sweet sorghum. Cellulosic biomass, derived from non-food sources such as trees and grasses is also being developed as a feedstock for ethanol production. Ethanol can be used as a fuel for vehicles in its pure form, but it is usually used as a gasoline additive to increase octane and improve vehicle emissions. Bioethanol is widely used in the USA and in Brazil. Biodiesel can be used as a fuel for vehicles in its pure form, but it is usually used as a diesel additive to reduce levels of particulates, carbon monoxide, and hydrocarbons from diesel-powered vehicles. Biodiesel is produced from oils or fats using transesterification and is the most common biofuel in Europe.

A bus fueled by biodiesel.

A solar vehicle is an electric vehicle powered completely or significantly by direct solar energy. Usually, photovoltaic (PV) cells contained in solar panels convert the sun's energy directly into electric energy. The term solar vehicle" usually implies that solar energy is used to power all or part of a vehicle's propulsion. Solar power may be also used to provide power for communications or controls or other auxiliary functions. Solar vehicles are not sold as practical day-to-day transportation devices at present, but are primarily demonstration vehicles and engineering exercises, often sponsored by government agencies. However, indirectly solar-charged vehicles are widespread and solar boats are available commercially.

History

Prior to the development of coal in the mid 19th century, nearly all energy used was renewable. Almost without a doubt the oldest known use of renewable energy, in the form of traditional biomass to fuel fires, dates from 790,000 years ago. Use of biomass for fire did not become commonplace until many hundreds of thousands of years later, sometime between 200,000 and 400,000 years ago. Probably the second oldest usage of renewable energy is harnessing the wind in order to drive ships over water. This practice can be traced back some 7000 years, to ships on the Nile. Moving into the time of recorded history, the primary sources of traditional renewable energy were human labor, animal power, water power, wind, in grain crushing windmills, and firewood, a traditional biomass. A graph of energy use in the United States up until 1900 shows oil and natural gas with about the same importance in 1900 as wind and solar played in 2010.

In the 1860s and ‹70s there were already fears that civilization would run out of fossil fuels and the need was felt for a better source. In 1873 Professor Augustine Mouchot wrote:

The time will arrive when the industry of Europe will cease to find those natural resources, so necessary for it. Petroleum springs and coal mines are not inexhaustible but are rapidly diminishing in many places. Will man, then, return to the power of water and wind? Or will he emigrate where the most powerful source of heat sends its rays to all? History will show what will come.

In 1885, Werner von Siemens, commenting on the discovery of the photovoltaic effect in the solid state, wrote:

In conclusion, I would say that however great the scientific importance of this discovery may be, its practical value will be no less obvious when we reflect that the supply of solar energy is both without limit and without cost, and that it will continue to pour down upon us for countless ages after all the coal deposits of the earth have been exhausted and forgotten.

Max Weber mentioned the end of fossil fuel in the concluding paragraphs of his Die protestantische Ethik und der Geist des Kapitalismus, published in 1905.

Development of solar engines continued until the outbreak of World War I. The importance of solar energy was recognized in a 1911 *Scientific American* article: “in the far distant future, natural fuels having been exhausted [solar power] will remain as the only means of existence of the human race”.

The theory of peak oil was published in 1956. In the 1970s environmentalists promoted the development of renewable energy both as a replacement for the eventual depletion of oil, as well as for an escape from dependence on oil, and the first electricity generating wind turbines appeared. Solar had long been used for heating and cooling, but solar panels were too costly to build solar farms until 1980.

The IEA 2014 World Energy Outlook projects a growth of renewable energy supply from 1,700 gigawatts in 2014 to 4,550 gigawatts in 2040. Fossil fuels received about $550 billion in subsidies in 2013, compared to $120 billion for all renewable energies.

Mainstream technologies

Wind Power

The 845 MW Shepherds Flat Wind Farm near Arlington, Oregon, USA.

Airflows can be used to run wind turbines. Modern utility-scale wind turbines range from around 600 kW to 5 MW of rated power, although turbines with rated output of 1.5–3 MW have become the most common for commercial use. The largest generator capacity of a single installed onshore wind turbine reached 7.5 MW in 2015. The power available from the wind is a function of the cube of the wind speed, so as wind speed increases, power output increases up to the maximum output for the particular turbine. Areas where winds are stronger and more constant, such as offshore and high altitude sites, are preferred locations for wind farms. Typically full load hours of wind turbines vary between 16 and 57 percent annually, but might be higher in particularly favorable offshore sites.

Wind-generated electricity met nearly 4% of global electricity demand in 2015, with nearly 63 GW of new wind power capacity installed. Wind energy was the leading source of new capacity in Europe, the US and Canada, and the second largest in China. In Denmark, wind energy met more than 40% of its electricity demand while Ireland, Portugal and Spain each met nearly 20%.

Globally, the long-term technical potential of wind energy is believed to be five times total current global energy production, or 40 times current electricity demand, assuming all practical barriers needed were overcome. This would require wind turbines to be installed over large areas, particularly in areas of higher wind resources, such as offshore. As offshore wind speeds average ~90% greater than that of land, so offshore resources can contribute substantially more energy than land stationed turbines. In 2014 global wind generation was 706 terawatt-hours or 3% of the worlds total electricity.

Hydropower

The Three Gorges Dam on the Yangtze River in China.

In 2015 hydropower generated 16.6% of the worlds total electricity and 70% of all renewable electricity. Since water is about 800 times denser than air, even a slow flowing stream of water, or moderate sea swell, can yield considerable amounts of energy. There are many forms of water energy:

- Historically hydroelectric power came from constructing large hydroelectric dams and reservoirs, which are still popular in third world countries. The largest of which is the Three Gorges Dam(2003) in China and the Itaipu Dam(1984) built by Brazil and Paraguay.
- Small hydro systems are hydroelectric power installations that typically produce up to 50 MW of power. They are often used on small rivers or as a low impact development on larger rivers. China is the largest producer of hydroelectricity in the world and has more than 45,000 small hydro installations.
- Run-of-the-river hydroelectricity plants derive kinetic energy from rivers without the creation of a large reservoir. This style of generation may still produce a large amount of electricity, such as the Chief Joseph Dam on the Columbia river in the United States.

Hydropower is produced in 150 countries, with the Asia-Pacific region generating 32 percent of global hydropower in 2010. For counties having the largest percentage of electricity from renewables, the top 50 are primarily hydroelectric. China is the largest hydroelectricity producer, with 721 terawatt-hours of production in 2010, representing around 17 percent of domestic electricity use. There are now three hydroelectricity stations larger than 10 GW: the Three Gorges Dam in China, Itaipu Dam across the Brazil/Paraguay border, and Guri Dam in Venezuela.

Wave power, which captures the energy of ocean surface waves, and tidal power, converting the energy of tides, are two forms of hydropower with future potential; however, they are not yet widely employed commercially. A demonstration project operated by the Ocean Renewable Power Company on the coast of Maine, and connected to the grid, harnesses tidal power from the Bay of Fundy, location of world's highest tidal flow. Ocean thermal energy conversion, which uses the temperature difference between cooler deep and warmer surface waters, has currently no economic feasibility.

Solar Energy

Satellite image of the 550-megawatt Topaz Solar Farm in California, USA

Solar energy, radiant light and heat from the sun, is harnessed using a range of ever-evolving technologies such as solar heating, photovoltaics, concentrated solar power (CSP), concentrator photovoltaics (CPV), solar architecture and artificial photosynthesis. Solar technologies are broadly characterized as either passive solar or active solar depending on the way they capture, convert and distribute solar energy. Passive solar techniques include orienting a building to the Sun, selecting materials with favorable thermal mass or light dispersing properties, and designing spaces that naturally circulate air. Active solar technologies encompass solar thermal energy, using solar collectors for heating, and solar power, converting sunlight into electricity either directly using photovoltaics (PV), or indirectly using concentrated solar power (CSP).

A photovoltaic system converts light into electrical direct current (DC) by taking advantage of the photoelectric effect. Solar PV has turned into a multi-billion, fast-growing industry, continues to improve its cost-effectiveness, and has the most potential of any renewable technologies together with CSP. Concentrated solar power (CSP) systems use lenses or mirrors and tracking systems to focus a large area of sunlight into a small beam. Commercial concentrated solar power plants were first developed in the 1980s. CSP-Stirling has by far the highest efficiency among all solar energy technologies.

In 2011, the International Energy Agency said that "the development of affordable, inexhaustible and clean solar energy technologies will have huge longer-term benefits. It will increase countries' energy security through reliance on an indigenous, inexhaustible and mostly import-independent resource, enhance sustainability, reduce pollution, lower the costs of mitigating climate change, and keep fossil fuel prices lower than otherwise. These advantages are global. Hence the additional costs of the incentives for early deployment should be considered learning investments; they must be wisely spent and need to be widely shared". In 2014 global solar generation was 186 terawatt-hours, slightly less than 1% of the worlds total grid electricity. Italy has the largest proportion of solar electricity in the world, in 2015 solar supplied 7.8% of electricity demand in Italy.

Geothermal Energy

Steam rising from the Nesjavellir Geothermal Power Station in Iceland.

High Temperature Geothermal energy is from thermal energy generated and stored in the Earth. Thermal energy is the energy that determines the temperature of matter. Earth's geothermal energy originates from the original formation of the planet and from radioactive decay of minerals (in currently uncertain but possibly roughly equalproportions). The geothermal gradient, which is the difference in temperature between the core of the planet and its surface, drives a continuous conduction of thermal energy in the form of heat from the core to the surface. The adjective *geothermal* originates from the Greek roots *geo*, meaning earth, and *thermos*, meaning heat.

The heat that is used for geothermal energy can be from deep within the Earth, all the way down to Earth's core – 4,000 miles (6,400 km) down. At the core, temperatures may reach over 9,000 °F (5,000 °C). Heat conducts from the core to surrounding rock. Extremely high temperature and pressure cause some rock to melt, which is commonly known as magma. Magma convects upward since it is lighter than the solid rock. This magma then heats rock and water in the crust, sometimes up to 700 °F (371 °C).

From hot springs, geothermal energy has been used for bathing since Paleolithic times and for space heating since ancient Roman times, but it is now better known for electricity generation.

Low Temperature Geothermal refers to the use of the outer crust of the earth as a Thermal Battery to facilitate Renewable thermal energy for heating and cooling buildings, and other refrigeration and industrial uses. In this form of Geothermal, a Geothermal Heat Pump and Ground-coupled heat exchanger are used together to move heat energy into the earth (for cooling) and out of the earth (for heating) on a varying seasonal basis. Low temperature Geothermal (generally referred to as "GHP") is an increasingly important renewable technology because it both reduces total annual energy loads associated with heating and cooling, and it also flattens the electric demand curve eliminating the extreme summer and winter peak electric supply requirements. Thus Low Temperature Geothermal/ GHP is becoming an increasing national priority with multiple tax credit support and focus as part of the ongoing movement toward Net Zero Energy. New York City has even just passed a law to require GHP anytime is shown to be economical with 20 year financing including the Socialized Cost of Carbon.

Bio Energy

Sugarcane plantation to produce ethanol in Brazil.

Biomass is biological material derived from living, or recently living organisms. It most often refers to plants or plant-derived materials which are specifically called lignocellulosic biomass. As an energy source, biomass can either be used directly via combustion to produce heat, or indirectly after converting it to various forms of biofuel. Conversion of biomass to biofuel can be achieved by different methods which are broadly classified into: *thermal*, *chemical*, and *biochemical* methods. Wood remains the largest biomass energy source today; examples include forest residues – such as dead trees, branches and tree stumps –, yard clippings, wood chips and even municipal solid waste. In the second sense, biomass includes plant or animal matter that can be converted into fibers or other industrial chemicals, including biofuels. Industrial biomass can be grown from numerous types of plants, including miscanthus, switchgrass, hemp, corn, poplar, willow, sorghum, sugarcane, bamboo, and a variety of tree species, ranging from eucalyptus to oil palm (palm oil).

A CHP power station using wood to supply 30,000 households in France.

Plant energy is produced by crops specifically grown for use as fuel that offer high biomass output per hectare with low input energy. Some examples of these plants are wheat, which typically yield 7.5–8 tonnes of grain per hectare, and straw, which typically yield 3.5–5 tonnes per hectare in the UK. The grain can be used for liquid transportation fuels while the straw can be burned to produce heat or electricity. Plant biomass can also be degraded from cellulose to glucose through a series of chemical treatments, and the resulting sugar can then be used as a first generation biofuel.

Biomass can be converted to other usable forms of energy like methane gas or transportation fuels like ethanol and biodiesel. Rotting garbage, and agricultural and human waste, all release methane gas – also called landfill gas or biogas. Crops, such as corn and sugarcane, can be fermented to produce the transportation fuel, ethanol. Biodiesel, another transportation fuel, can be produced from left-over food products like vegetable oils and animal fats. Also, biomass to liquids (BTLs) and cellulosic ethanol are still under research. There is a great deal of research involving algal fuel or algae-derived biomass due to the fact that it's a non-food resource and can be produced at rates 5 to 10 times those of other types of land-based agriculture, such as corn and soy. Once harvested, it can be fermented to produce biofuels such as ethanol, butanol, and methane, as well as biodiesel and hydrogen. The biomass used for electricity generation varies by region. Forest by-products, such as wood residues, are common in the United States. Agricultural waste is common in Mauritius (sugar cane residue) and Southeast Asia (rice husks). Animal husbandry residues, such as poultry litter, are common in the United Kingdom.

Biofuels include a wide range of fuels which are derived from biomass. The term covers solid, liquid, and gaseous fuels.Liquid biofuels include bioalcohols, such as bioethanol, and oils, such as biodiesel. Gaseous biofuels include biogas, landfill gas and synthetic gas. Bioethanol is an alcohol made by fermenting the sugar components of plant materials and it is made mostly from sugar and starch crops. These include maize, sugarcane and, more recently, sweet sorghum. The latter crop is particularly suitable for growing in dryland conditions, and is being investigated by International Crops Research Institute for the Semi-Arid Tropics for its potential to provide fuel, along with food and animal feed, in arid parts of Asia and Africa.

With advanced technology being developed, cellulosic biomass, such as trees and grasses, are also used as feedstocks for ethanol production. Ethanol can be used as a fuel for vehicles in its pure form, but it is usually used as a gasoline additive to increase octane and improve vehicle emissions. Bioethanol is widely used in the United States and in Brazil. The energy costs for producing bio-ethanol are almost equal to, the energy yields from bio-ethanol. However, according to the European Environment Agency, biofuels do not address global warming concerns. Biodiesel is made from vegetable oils, animal fats or recycled greases. It can be used as a fuel for vehicles in its pure form, or more commonly as a diesel additive to reduce levels of particulates, carbon monoxide, and hydrocarbons from diesel-powered vehicles. Biodiesel is produced from oils or fats using transesterification and is the most common biofuel in Europe. Biofuels provided 2.7% of the world's transport fuel in 2010.

Biomass, biogas and biofuels are burned to produce heat/power and in doing so harm the environment. Pollutants such as sulphurous oxides (SO_x), nitrous oxides (NO_x), and particulate matter (PM) are produced from the combustion of biomass; the World Health Organisation estimates that 7 million premature deaths are caused each year by air pollution. Biomass combustion is a major contributor. The life cycle of the plants is sustainable, the lives of people less so.

Energy Storage

Energy storage is a collection of methods used to store electrical energy on an electrical power grid, or off it. Electrical energy is stored during times when production (especially from intermittent power plants such as renewable electricity sources such as wind power, tidal power, solar power) exceeds consumption, and returned to the grid when production falls below consumption. Water pumped into a hydroelectric dam is the largest form of power storage.

Market and Industry Trends

Growth of Renewables

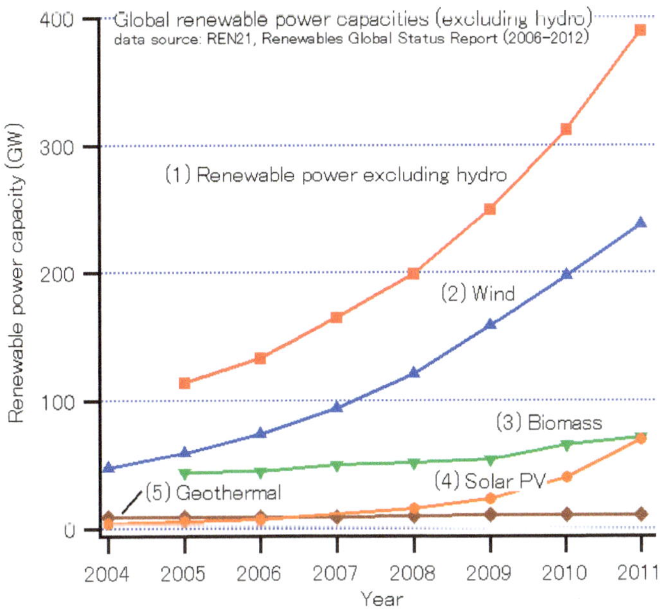

Global growth of renewables through to 2011

From the end of 2004, worldwide renewable energy capacity grew at rates of 10–60% annually for many technologies. In 2015 global investment in renewables rose 5% to \$285.9 billion, breaking the previous record of \$278.5 billion in 2011. 2015 was also the first year that saw renewables, excluding large hydro, account for the majority of all new power capacity (134GW, making up 53.6% of the total). Of the renewables total, wind accounted for 72GW and solar photovoltaics 56GW; both record-breaking numbers and sharply up from 2014 figures (49GW and 45GW respectively). In financial terms, solar made up 56% of total new investment and wind accounted for 38%.

Projections vary. The EIA has predicted that almost two thirds of net additions to power capacity will come from renewables by 2020 due to the combined policy benefits of local pollution, decarbonisation and energy diversification. Some studies have set out roadmaps to power 100% of the world's energy with wind, hydroelectric and solar by the year 2030.

According to a 2011 projection by the International Energy Agency, solar power generators may produce most of the world's electricity within 50 years, reducing the emissions of greenhouse gases that harm the environment. Cedric Philibert, senior analyst in the renewable energy division at the IEA said: "Photovoltaic and solar-thermal plants may meet most of the world's demand for electricity by 2060 – and half of all energy needs – with wind, hydropower and biomass plants supplying much of the remaining generation". "Photovoltaic and concentrated solar power together can become the major source of electricity", Philibert said.

In 2014 global wind power capacity expanded 16% to 369,553 MW. Yearly wind energy production is also growing rapidly and has reached around 4% of worldwide electricity usage, 11.4% in the EU, and it is widely used in Asia, and the United States. In 2015, worldwide installed photovoltaics capacity increased to 227 gigawatts (GW), sufficient to supply 1 percent of global electricity demands. Solar thermal energy stations operate in the USA and Spain, and as of 2016, the largest of these is the 392 MW Ivanpah Solar Electric Generating System in California. The world's largest geothermal power installation is The Geysers in California, with a rated capacity of 750 MW. Brazil has one of the largest renewable energy programs in the world, involving production of ethanol fuel from sugar cane, and ethanol now provides 18% of the country's automotive fuel. Ethanol fuel is also widely available in the USA.

Selected renewable energy global indicators	**2008**	**2009**	**2010**	**2011**	**2012**	**2013**	**2014**	**2015**
Investment in new renewable capacity (annual) (10^9 USD)	182	178	237	279	256	232	270	285
Renewables power capacity (existing) (GWe)	1,140	1,230	1,320	1,360	1,470	1,578	1,712	1,849
Hydropower capacity (existing) (GWe)	885	915	945	970	990	1,018	1,055	1,064
Wind power capacity (existing) (GWe)	121	159	198	238	283	319	370	433
Solar PV capacity (grid-connected) (GWe)	16	23	40	70	100	138	177	227
Solar hot water capacity (existing) (GWth)	130	160	185	232	255	373	406	435
Ethanol production (annual) (10^9 litres)	67	76	86	86	83	87	94	98
Biodiesel production (annual) (10^9 litres)	12	17.8	18.5	21.4	22.5	26	29.7	30
Countries with policy targets for renewable energy use	79	89	98	118	138	144	164	173
Source: The Renewable Energy Policy Network for the 21st Century (REN21)–Global Status Report								

Economic Trends

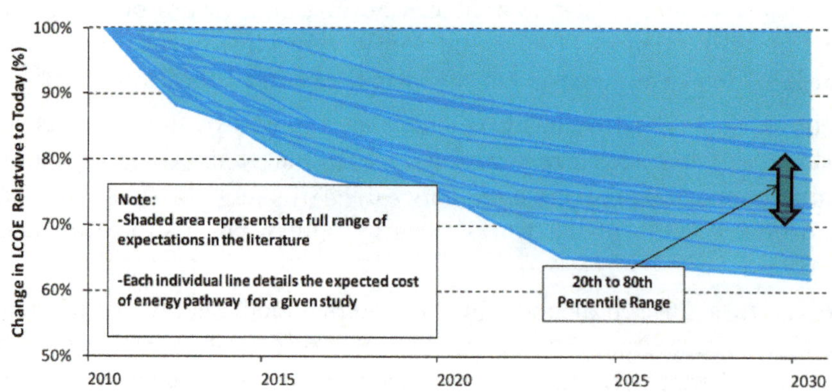

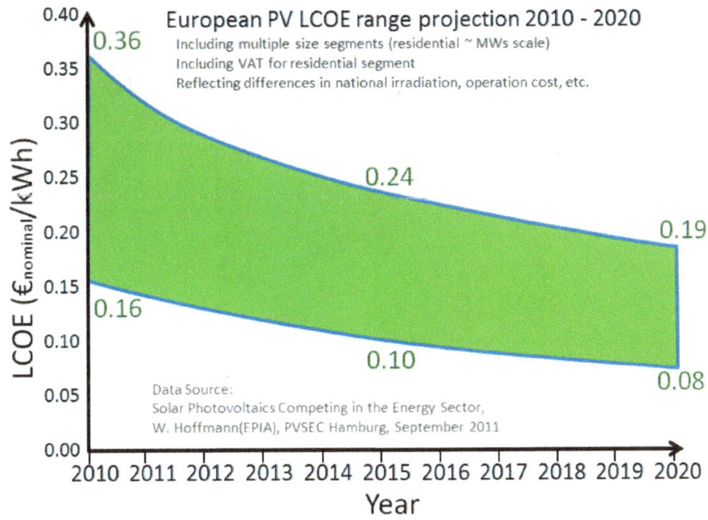

Projection of levelized cost for wind in the U.S. (left) and solar power in Europe.

Renewable energy technologies are getting cheaper, through technological change and through the benefits of mass production and market competition. A 2011 IEA report said: A portfolio of renewable energy technologies is becoming cost-competitive in an increasingly broad range of circumstances, in some cases providing investment opportunities without the need for specific economic support," and added that cost reductions in critical technologies, such as wind and solar, are set to continue."

Hydro-electricity and geothermal electricity produced at favourable sites are now the cheapest way to generate electricity. Renewable energy costs continue to drop, and the levelised cost of electricity (LCOE) is declining for wind power, solar photovoltaic (PV), concentrated solar power (CSP) and some biomass technologies. Renewable energy is also the most economic solution for new grid-connected capacity in areas with good resources. As the cost of renewable power falls, the scope of economically viable applications increases. Renewable technologies are now often the most economic solution for new generating capacity. Where oil-fired generation is the predominant power generation source (e.g. on islands, off-grid and in some countries) a lower-cost renewable solution almost always exists today". A series of studies by the US National Renewable Energy Laboratory modeled the "grid in the Western US under a number of different scenarios where intermittent renewables accounted for 33 percent of the total power." In the models, inefficiencies in cycling the fossil fuel plants to compensate for the variation in solar and wind energy resulted in an additional cost of "between $0.47 and $1.28 to each MegaWatt hour generated"; however, the savings in the cost of the fuels saved adds up to $7 billion, meaning the added costs are, at most, two percent of the savings."

Hydroelectricity

Only 25% of the worlds estimated hydroelectric potential of 14,000 TWh/year has been developed, with Africa, Asia and Latin America having the greatest potential. The Three

Gorges Dam in Hubei, China, has the world's largest instantaneous generating capacity (22,500 MW), with the Itaipu Dam in Brazil/Paraguay in second place (14,000 MW). The Three Gorges Dam is operated jointly with the much smaller Gezhouba Dam (3,115 MW). As of 2012, the total generating capacity of this two-dam complex is 25,615 MW. In 2008, this complex generated 98 TWh of electricity (81 TWh from the Three Gorges Dam and 17 TWh from the Gezhouba Dam), which is 3% more power in one year than the 95 TWh generated by Itaipu in 2008.

Wind Power Development

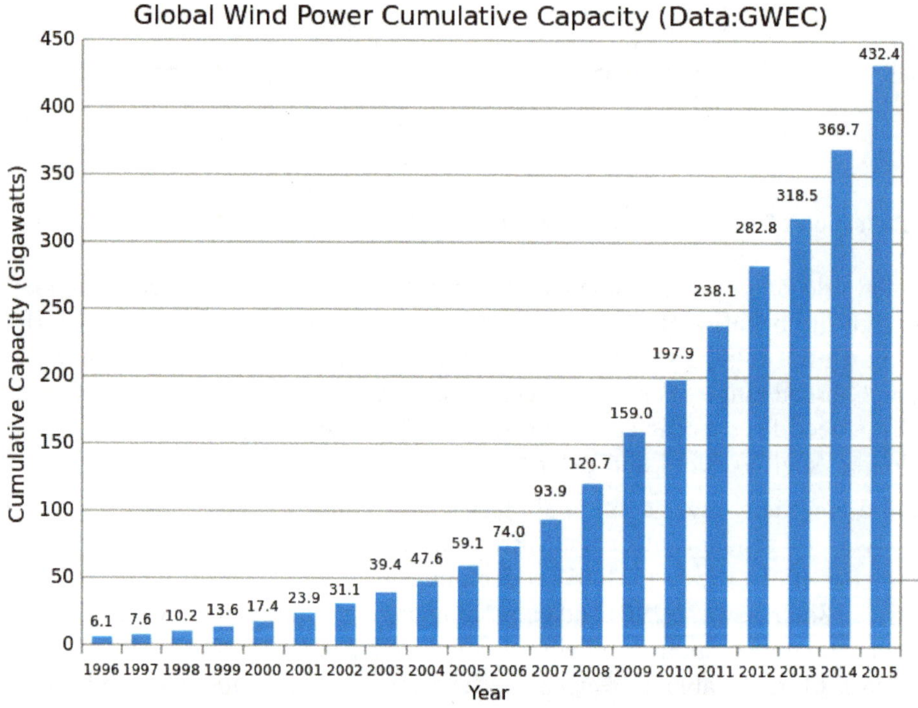

Worldwide growth of wind capacity (1996–2014).

Wind power is widely used in Europe, China, and the United States. From 2004 to 2014, worldwide installed capacity of wind power has been growing from 47 GW to 369 GW—a more than sevenfold increase within 10 years with 2014 breaking a new record in global installations (51 GW). As of the end of 2014, China, the United States and Germany combined accounted for half of total global capacity.Several other countries have achieved relatively high levels of wind power penetration, such as 21% of stationary electricity production in Denmark, 18% in Portugal, 16% in Spain, and 14% in Ireland in 2010 and have since continued to expand their installed capacity.More than 80 countries around the world are using wind power on a commercial basis.

Four offshore wind farms are in the Thames Estuary area: Kentish Flats, Gunfleet Sands, Thanet and London Array. The latter is the largest in the world as of April 2013.

- Offshore wind power

As of 2014, offshore wind power amounted to 8,771 megawatt of global installed capacity. Although offshore capacity doubled within three years (from 4,117 MW in 2011), it accounted for only 2.3% of the total wind power capacity. The United Kingdom is the undisputed leader of offshore power with half of the world's installed capacity ahead of Denmark, Germany, Belgium and China.

- List of offshore and onshore wind farms

As of 2012, the Alta Wind Energy Center (California, 1,020 MW) is the world's largest wind farm. The London Array (630 MW) is the largest offshore wind farm in the world. The United Kingdom is the world's leading generator of offshore wind power, followed by Denmark. There are several large offshore wind farms operational and under construction and these include Anholt (400 MW), BARD (400 MW), Clyde (548 MW), Fântânele-Cogealac (600 MW), Greater Gabbard (500 MW), Lincs (270 MW), London Array (630 MW), Lower Snake River (343 MW), Macarthur (420 MW), Shepherds Flat (845 MW), and the Sheringham Shoal (317 MW).

Solar Thermal

The 377 MW Ivanpah Solar Electric Generating System with all three towers under load, Feb 2014. Taken from I-15.

Solar Towers of the PS10 and PS20solar thermal plants in Spain.

The United States conducted much early research in photovoltaics and concentrated solar power. The U.S. is among the top countries in the world in electricity generated by the Sun and several of the world's largest utility-scale installations are located in the desert Southwest.

The oldest solar thermal power plant in the world is the 354 megawatt (MW) SEGS thermal power plant, in California. The Ivanpah Solar Electric Generating System is a solar thermal power project in the California Mojave Desert, 40 miles (64 km) southwest of Las Vegas, with a gross capacity of 377 MW. The 280 MW Solana Generating Station is a solar power plant near Gila Bend, Arizona, about 70 miles (110 km) southwest of Phoenix, completed in 2013. When commissioned it was the largest parabolic trough plant in the world and the first U.S. solar plant with molten salt thermal energy storage.

The solar thermal power industry is growing rapidly with 1.3 GW under construction in 2012 and more planned. Spain is the epicenter of solar thermal power development with 873 MW under construction, and a further 271 MW under development. In the United States, 5,600 MW of solar thermal power projects have been announced.Several power plants have been constructed in the Mojave Desert, Southwestern United States. The Ivanpah Solar Power Facility being the most recent. In developing countries, three World Bank projects for integrated solar thermal/combined-cycle gas-turbine power plants in Egypt, Mexico, and Morocco have been approved.

Photovoltaic Development

Worldwide growth of PV capacity grouped by region in MW (2006–2014)

Photovoltaics (PV) uses solar cells assembled into solar panels to convert sunlight into electricity. It's a fast-growing technology doubling its worldwide installed capacity every couple of years. PV systems range from small, residential and commercial rooftop or building integrated installations, to large utility-scale photovoltaic power station. The predominant PV technology is crystalline silicon, while thin-film solar cell technology accounts for about 10 percent of global photovoltaic deployment. In recent years, PV technology has improved its electricity generating efficiency, reduced the installation cost per watt as well as its energy payback time, and has reached grid parity in at least 30 different markets by 2014. Financial institutions are predicting a second solar "gold rush" in the near future.

At the end of 2014, worldwide PV capacity reached at least 177,000 megawatts. Photovoltaics grew fastest in China, followed by Japan and the United States, while Germany remains the world's largest overall producer of photovoltaic power, contributing about 7.0 percent to the overall electricity generation. Italy meets 7.9 percent of its electricity demands with photovoltaic power—the highest share worldwide. For 2015, global cumulative capacity is forecasted to increase by more than 50 gigawatts (GW). By 2018, worldwide capacity is projected to reach as much as 430 gigawatts. This corresponds to a tripling within five years. Solar power is forecasted to become the world's largest source of electricity by 2050, with solar photovoltaics and concentrated solar power contributing 16% and 11%, respectively. This requires an increase of installed PV capacity to 4,600 GW, of which more than half is expected to be deployed in China and India.

Photovoltaic Power Stations

Solar panels at the 550 MW Topaz Solar Farm.

Nellis Solar Power Plant, photovoltaic power plant in Nevada, USA.

Commercial concentrated solar power plants were first developed in the 1980s. As the cost of solar electricity has fallen, the number of grid-connected solar PV systems has grown into the millions and utility-scale solar power stations with hundreds of megawatts are being built. Solar PV is rapidly becoming an inexpensive, low-carbon technology to harness renewable energy from the Sun.

Many solar photovoltaic power stations have been built, mainly in Europe, China and the USA. The 579 MW Solar Star, in the United States, is the world's largest PV power station.

Many of these plants are integrated with agriculture and some use tracking systems that follow the sun's daily path across the sky to generate more electricity than fixed-mounted systems. There are no fuel costs or emissions during operation of the power stations.

However, when it comes to renewable energy systems and PV, it is not just large systems that matter. Building-integrated photovoltaics or "onsite" PV systems use existing land and structures and generate power close to where it is consumed.

Biofuel Development

Brazil produces bioethanol made from sugarcane available throughout the country. A typical gas station with dual fuel service is marked "A" for alcohol (ethanol) and "G" for gasoline.

Biofuels provided 3% of the world's transport fuel in 2010. Mandates for blending biofuels exist in 31 countries at the national level and in 29 states/provinces.According to the International Energy Agency, biofuels have the potential to meet more than a quarter of world demand for transportation fuels by 2050.

Since the 1970s, Brazil has had an ethanol fuel program which has allowed the country to become the world's second largest producer of ethanol (after the United States) and the world's largest exporter. Brazil's ethanol fuel program uses modern equipment and cheap sugarcane as feedstock, and the residual cane-waste (bagasse) is used to produce heat and power. There are no longer light vehicles in Brazil running on pure gasoline. By the end of 2008 there were 35,000 filling stations throughout Brazil with at least one ethanol pump. Unfortunately, Operation Car Wash has seriously eroded public trust in oil companies and has implicated several high ranking Brazilian officials.

Nearly all the gasoline sold in the United States today is mixed with 10% ethanol,and motor vehicle manufacturers already produce vehicles designed to run on much higher ethanol blends. Ford, Daimler AG, and GM are among the automobile companies that sell "flexible-fuel" cars, trucks, and minivans that can use gasoline and ethanol blends ranging from pure gasoline up to 85% ethanol. By mid-2006, there were approximately 6 million ethanol compatible vehicles on U.S. roads.

Geothermal Development

Geothermal plant at The Geysers, California, USA.

Geothermal power is cost effective, reliable, sustainable, and environmentally friendly, but has historically been limited to areas near tectonic plate boundaries. Recent technological advances have expanded the range and size of viable resources, especially for applications such as home heating, opening a potential for widespread exploitation. Geothermal wells release greenhouse gases trapped deep within the earth, but these emissions are much lower per energy unit than those of fossil fuels. As a result, geothermal power has the potential to help mitigate global warming if widely deployed in place of fossil fuels.

The International Geothermal Association (IGA) has reported that 10,715 MW of geothermal power in 24 countries is online, which is expected to generate 67,246 GWh of electricity in 2010. This represents a 20% increase in geothermal power online capacity since 2005. IGA projects this will grow to 18,500 MW by 2015, due to the large number of projects presently under consideration, often in areas previously assumed to have little exploitable resource.

In 2010, the United States led the world in geothermal electricity production with 3,086 MW of installed capacity from 77 power plants; the largest group of geothermal power plants in the world is located at The Geysers, a geothermal field in California. The Philippines follows the US as the second highest producer of geothermal power in the world, with 1,904 MW of capacity online; geothermal power makes up approximately 18% of the country's electricity generation.

Developing Countries

Solar cookers use sunlight as energy source for outdoor cooking.

Renewable energy technology has sometimes been seen as a costly luxury item by critics, and affordable only in the affluent developed world. This erroneous view has persisted for many years, but 2015 was the first year when investment in non-hydro renewables, was higher in developing countries, with $156 billion invested, mainly in China, India, and Brazil.

Renewable energy can be particularly suitable for developing countries. In rural and remote areas, transmission and distribution of energy generated from fossil fuels can be difficult and expensive. Producing renewable energy locally can offer a viable alternative.

Technology advances are opening up a huge new market for solar power: the approximately 1.3 billion people around the world who don't have access to grid electricity. Even though they are typically very poor, these people have to pay far more for lighting than people in rich countries because they use inefficient kerosene lamps. Solar power costs half as much as lighting with kerosene. As of 2010, an estimated 3 million households get power from small solar PV systems. Kenya is the world leader in the number of solar power systems installed per capita. More than 30,000 very small solar panels, each producing 12 to 30 watts, are sold in Kenya annually. Some Small Island Developing States (SIDS) are also turning to solar power to reduce their costs and increase their sustainability.

Micro-hydro configured into mini-grids also provide power. Over 44 million households use biogas made in household-scale digesters for lighting and/or cooking, and more than 166 million households rely on a new generation of more-efficient biomass cookstoves. Clean liquid fuel sourced from renewable feedstocks are used for cooking and lighting in energy-poor areas of the developing world. Alcohol fuels (ethanol and methanol) can be produced sustainably from non-food sugary, starchy, and cellulostic feedstocks. Project Gaia, Inc. and CleanStar Mozambique are implementing clean cooking programs with liquid ethanol stoves in Ethiopia, Kenya, Nigeria and Mozambique.

Renewable energy projects in many developing countries have demonstrated that renewable energy can directly contribute to poverty reduction by providing the energy needed for creating businesses and employment. Renewable energy technologies can also make indirect contributions to alleviating poverty by providing energy for cooking, space heating, and lighting. Renewable energy can also contribute to education, by providing electricity to schools.

Industry and Policy Trends

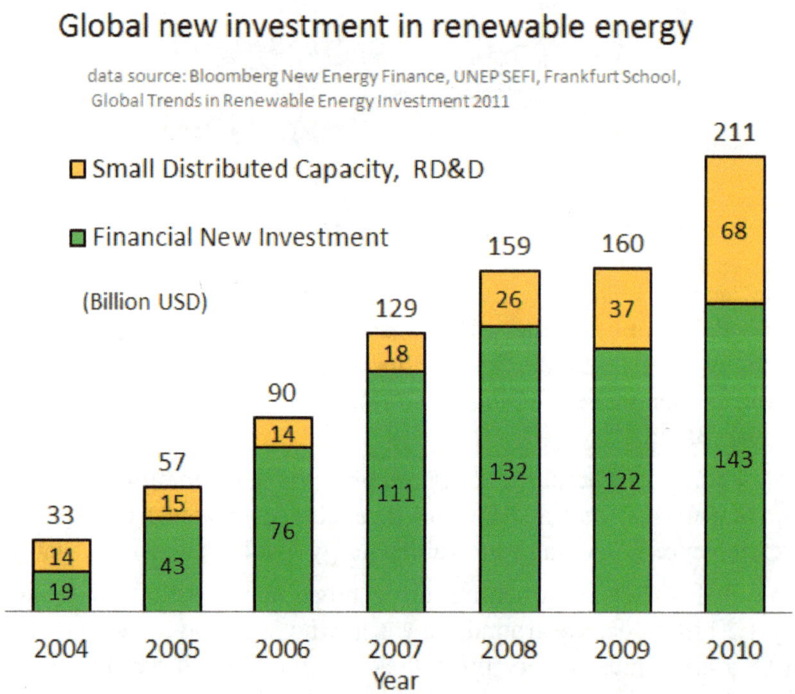

Global New Investments in Renewable Energy

U.S. President Barack Obama's American Recovery and Reinvestment Act of 2009 includes more than $70 billion in direct spending and tax credits for clean energy and associated transportation programs. Leading renewable energy companies include First Solar, Gamesa, GE Energy, Hanwha Q Cells, Sharp Solar, Siemens, SunOpta, Suntech Power, and Vestas.

Many national, state, and local governments have also created green banks. A green bank is a quasi-public financial institution that uses public capital to leverage private investment in clean energy technologies. Green banks use a variety of financial tools to bridge market gaps that hinder the deployment of clean energy.

The military has also focused on the use of renewable fuels for military vehicles. Unlike fossil fuels, renewable fuels can be produced in any country, creating a strategic advantage. The US military has already committed itself to have 50% of its energy consumption come from alternative sources.

The International Renewable Energy Agency (IRENA) is an intergovernmental organization for promoting the adoption of renewable energy worldwide. It aims to provide concrete policy advice and facilitate capacity building and technology transfer. IRENA was formed on 26 January 2009, by 75 countries signing the charter of IRENA. As of March 2010, IRENA has 143 member states who all are considered as founding members, of which 14 have also ratified the statute.

As of 2011, 119 countries have some form of national renewable energy policy target or renewable support policy. National targets now exist in at least 98 countries. There is also a wide range of policies at state/provincial and local levels.

United Nations' Secretary-General Ban Ki-moon has said that renewable energy has the ability to lift the poorest nations to new levels of prosperity. In October 2011, he "announced the creation of a high-level group to drum up support for energy access, energy efficiency and greater use of renewable energy. The group is to be co-chaired by Kandeh Yumkella, the chair of UN Energy and director general of the UN Industrial Development Organisation, and Charles Holliday, chairman of Bank of America".

100% Renewable Energy

The incentive to use 100% renewable energy, for electricity, transport, or even total primary energy supply globally, has been motivated by global warming and other ecological as well as economic concerns. The Intergovernmental Panel on Climate Change has said that there are few fundamental technological limits to integrating a portfolio of renewable energy technologies to meet most of total global energy demand. Renewable energy use has grown much faster than even advocates anticipated. At the national level, at least 30 nations around the world already have renewable energy contributing more than 20% of energy supply. Also, Professors S. Pacala and Robert H. Socolow have developed a series of "stabilization wedges" that can allow us to maintain our quality of life while avoiding catastrophic climate change, and renewable energy sources," in aggregate, constitute the largest number of their wedges".

Using 100% renewable energy was first suggested in a Science paper published in 1975 by Danish physicist Bent Sørensen. It was followed by several other proposals, until in 1998 the first detailed analysis of scenarios with very high shares of renewables were published. These were followed by the first detailed 100% scenarios. In 2006 a PhD thesis was published by Czisch in which it was shown that in a 100% renewable scenario energy supply could match demand in every hour of the year in Europa and North Africa. In the same year Danish Energy professor Henrik Lund published a first paperin which he addresses the optimal combination of renewables, which was followed by several other papers on the transition to 100% renewable energy in Denmark. Since then Lund has been publishing several papers on 100% renewable energy. After 2009 publications began to rise steeply, covering 100% scenarios for countries in Europa, America, Australia and other parts of the world.

In 2011 Mark Z. Jacobson, professor of civil and environmental engineering at Stanford University, and Mark Delucchi published a study on 100% renewable global energy supply in the journal Energy Policy. They found producing all new energy with wind power, solar power, and hydropower by 2030 is feasible and existing energy supply arrangements could be replaced by 2050. Barriers to implementing the renewable energy plan are seen to be "primarily social and political, not technological or economic". They also found that energy costs with a wind, solar, water system should be similar to today's energy costs.

Similarly, in the United States, the independent National Research Council has noted that sufficient domestic renewable resources exist to allow renewable electricity to play a significant role in future electricity generation and thus help confront issues related to climate change, energy security, and the escalation of energy costs ... Renewable energy is an attractive option because renewable resources available in the United States, taken collectively, can supply significantly greater amounts of electricity than the total current or projected domestic demand." .

The most significant barriers to the widespread implementation of large-scale renewable energy and low carbon energy strategies are primarily political and not technological. According to the 2013 *Post Carbon Pathways* report, which reviewed many international studies, the key roadblocks are: climate change denial, the fossil fuels lobby, political inaction, unsustainable energy consumption, outdated energy infrastructure, and financial constraints.

Emerging Technologies

Other renewable energy technologies are still under development, and include cellulosic ethanol, hot-dry-rock geothermal power, and marine energy. These technologies are not yet widely demonstrated or have limited commercialization. Many are on the horizon and may have potential comparable to other renewable energy technologies, but still depend on attracting sufficient attention and research, development and demonstration (RD&D) funding.

There are numerous organizations within the academic, federal, and commercial sectors conducting large scale advanced research in the field of renewable energy. This research spans several areas of focus across the renewable energy spectrum. Most of the research is targeted at improving efficiency and increasing overall energy yields. Multiple federally supported research organizations have focused on renewable energy in recent years. Two of the most prominent of these labs are Sandia National Laboratories and the National Renewable Energy Laboratory (NREL), both of which are funded by the United States Department of Energy and supported by various corporate partners. Sandia has a total budget of $2.4 billionwhile NREL has a budget of $375 million.

- Enhanced geothermal system

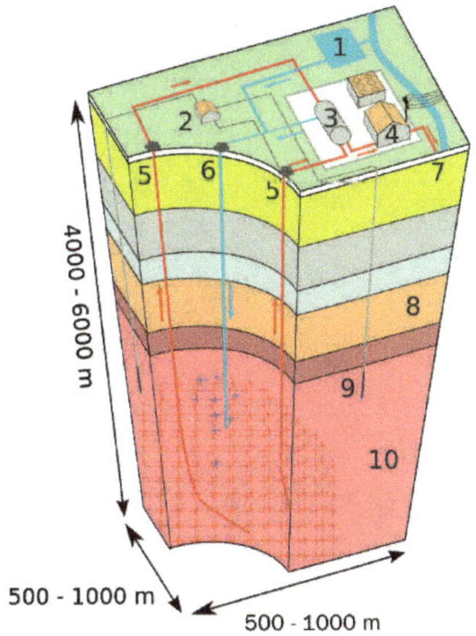

Enhanced geothermal system .

Enhanced geothermal systems (EGS) are a new type of geothermal power technologies that do not require natural convective hydrothermal resources. The vast majority of geothermal energy within drilling reach is in dry and non-porous rock. EGS technologies "enhance" and/or create geothermal resources in this "hot dry rock (HDR)" through hydraulic stimulation. EGS and HDR technologies, like hydrothermal geothermal, are expected to be baseload resources which produce power 24 hours a day like a fossil plant. Distinct from hydrothermal, HDR and EGS may be feasible anywhere in the world, depending on the economic limits of drill depth. Good locations are over deep granite covered by a thick (3–5 km) layer of insulating sediments which slow heat loss. There are HDR and EGS systems currently being developed and tested in France, Australia, Japan, Germany, the U.S. and Switzerland. The largest EGS project in the world is a 25 megawatt demonstration plant currently being developed in the Cooper Basin, Australia. The Cooper Basin has the potential to generate 5,000–10,000 MW.

- Cellulosic ethanol

Several refineries that can process biomass and turn it into ethanol are built by companies such as Iogen, POET, and Abengoa, while other companies such as the Verenium Corporation, Novozymes, and Dyadic International are producing enzymes which could enable future commercialization. The shift from food crop feedstocks to waste residues and native grasses offers significant opportunities for a range of players, from farmers to biotechnology firms, and from project developers to investors.

- Marine energy

Rance Tidal Power Station, France.

Marine energy (also sometimes referred to as ocean energy) refers to the energy carried by ocean waves, tides, salinity, and ocean temperature differences. The movement of water in the world's oceans creates a vast store of kinetic energy, or energy in motion. This energy can be harnessed to generate electricity to power homes, transport and industries. The term marine energy encompasses both wave power – power from surface waves, and tidal power – obtained from the kinetic energy of large bodies of moving water. Reverse electrodialysis (RED) is a technology for generating electricity by mixing fresh river water and salty sea water in large power cells designed for this purpose; as of 2016 it is being tested at a small scale (50 kW). Offshore wind power is not a form of marine energy, as wind power is derived from the wind, even if the wind turbines are placed over water. The oceans have a tremendous amount of energy and are close to many if not most concentrated populations. Ocean energy has the potential of providing a substantial amount of new renewable energy around the world.

- Experimental solar power

Concentrated photovoltaics (CPV) systems employ sunlight concentrated onto photovoltaic surfaces for the purpose of electricity generation. Thermoelectric, or thermovoltaic" devices convert a temperature difference between dissimilar materials into an electric current.

- Floating solar arrays

Floating solar arrays are PV systems that float on the surface of drinking water reservoirs, quarry lakes, irrigation canals or remediation and tailing ponds. A small number of such systems exist in France, India, Japan, South Korea, the United Kingdom, Singapore and the United States. The systems are said to have advantages over photovoltaics on land. The cost of land is more expensive, and there are fewer rules and regulations for structures built on bodies of water not used for recreation. Unlike most land-based solar plants, floating arrays can be unobtrusive because they are hidden from public view. They achieve higher efficiencies than PV panels on land, because water cools the panels. The panels have a special coating to prevent rust or corrosion. In May 2008, the Far Niente Winery in Oakville, California, pioneered the world's first floatovoltaic system by installing 994 solar PV modules with a total capacity of 477 kW onto 130 pontoons and floating them on the winery's irrigation pond. Utility-scale floating PV farms are starting to be built. Kyocera will develop the world's largest, a 13.4 MW farm on the reservoir above Yamakura Dam in Chiba Prefecture using 50,000 solar panels. Salt-water resistant floating farms are also being constructed for ocean use. The largest so far announced floatovoltaic project is a 350 MW power station in the Amazon region of Brazil.

- Solar-assisted heat pump

A heat pump is a device that provides heat energy from a source of heat to a destination called a "heat sink". Heat pumps are designed to move thermal energy opposite to the direction of spontaneous heat flow by absorbing heat from a cold space and releasing it to a warmer one. A solar-assisted heat pump represents the integration of a heat pump and thermal solar panels in a single integrated system. Typically these two technologies are used separately (or only placing them in parallel) to produce hot water. In this system the solar thermal panel performs the function of the low temperature heat source and the heat produced is used to feed the heat pump's evaporator. The goal of this system is to get high COP and then produce energy in a more efficient and less expensive way.

It is possible to use any type of solar thermal panel (sheet and tubes, roll-bond, heat pipe, thermal plates) or hybrid (mono/polycrystalline, thin film) in combination with the heat pump. The use of a hybrid panel is preferable because it allows to cover a part of the electricity demand of the heat pump and reduce the power consumption and consequently the variable costs of the system.

- Artificial photosynthesis

Artificial photosynthesis uses techniques including nanotechnology to store solar electromagnetic energy in chemical bonds by splitting water to produce hydrogen and then using carbon dioxide to make methanol. Researchers in this field are striving to design molecular mimics of photosynthesis that utilize a wider region of the solar spectrum, employ catalytic systems made from abundant, inexpensive materials that are robust, readily repaired, non-toxic, stable in a variety of environmental conditions and perform more efficiently allowing a greater proportion of photon energy to end up in the storage compounds, i.e., carbohydrates (rather than building and sustaining living cells). However, prominent research faces hurdles, Sun Catalytix a MIT spin-off stopped scaling up their prototype fuel-cell in 2012, because it offers few savings over other ways to make hydrogen from sunlight.

- Algae fuels

Producing liquid fuels from oil-rich varieties of algae is an ongoing research topic. Various microalgae grown in open or closed systems are being tried including some system that can be set up in brownfield and desert lands.

- Solar aircraft

In 2016, *Solar Impulse 2* was the first solar-powered aircraft to complete circumnavigation of the world.

An electric aircraft is an aircraft that runs on electric motors rather than internal combustion engines, with electricity coming from fuel cells, solar cells, ultracapacitors, power beaming, or batteries.

Currently, flying manned electric aircraft are mostly experimental demonstrators, though many small unmanned aerial vehicles are powered by batteries. Electrically powered model aircraft have been flown since the 1970s, with one report in 1957. The first man-carrying electrically powered flights were made in 1973. Between 2015-2016, a manned, solar-powered plane, Solar Impulse 2, completed a circumnavigation of the Earth.

- Solar updraft tower

The Solar updraft tower is a renewable-energy power plant for generating electricity from low temperature solar heat. Sunshine heats the air beneath a very wide greenhouse-like roofed collector structure surrounding the central base of a very tall chimney tower. The resulting convection causes a hot air updraft in the tower by the chimney effect. This airflow drives wind turbines placed in the chimney updraft or around the chimney base to produce electricity. Plans for scaled-up versions of demonstration models will allow significant power generation, and may allow development of other applications, such as water extraction or distillation, and agriculture or horticulture. A more advanced version of a similarly themed technology is the Vortex engine which aims to replace large physical chimneys with a vortex of air created by a shorter, less-expensive structure.

- Space-based solar power

For either photovoltaic or thermal systems, one option is to loft them into space, particularly Geosynchronous orbit. To be competitive with Earth-based solar power systems, the specific mass (kg/kW) times the cost to loft mass plus the cost of the parts needs to be $2400 or less. I.e., for a parts cost plus rectenna of $1100/kW, the product of the $/kg and kg/kW must be $1300/kW or less. Thus for 6.5 kg/kW, the transport cost cannot exceed $200/kg. While that will require a 100 to one reduction, SpaceX is targeting a ten to one reduction, Reaction Engines may make a 100 to one reduction possible.

- Carbon-neutral and negative fuels

Carbon-neutral fuels are synthetic fuels (including methane, gasoline, diesel fuel, jet fuel or ammonia) produced by hydrogenating waste carbon dioxide recycled from power plant flue-gas emissions, recovered from automotive exhaust gas, or derived from carbonic acid in seawater. Such fuels are considered carbon-neutral because they do not result in a net increase in atmospheric greenhouse gases.

Carbon-neutral fuels offer relatively low cost energy storage, alleviating the problems of wind and solar variability, and they enable distribution of wind, water, and solar power through existing natural gas pipelines. Nighttime wind power is considered the most economical form of electrical power with which to synthesize fuel, because the load curve for electricity peaks sharply during the warmest hours of the day, but wind tends to blow slightly more at night than during the day, so, the price of nighttime wind power is often much less expensive than any alternative. Germany has built a 250 kilowatt synthetic methane plant which they are scaling up to 10 megawatts.

The George Olah carbon dioxide recycling plant in Grindavík, Iceland has been producing 2 million liters of methanol transportation fuel per year from flue exhaust of the Svartsengi Power Station since 2011. It has the capacity to produce 5 million liters per year.

Debate

Renewable electricity production, from sources such as wind power and solar power, is sometimes criticized for being variable or intermittent, but is not true for concentrated solar, geothermal and biofuels, that have continuity. In any case, the International Energy Agency has stated that deployment of renewable technologies usually increases the diversity of electricity sources and, through local generation, contributes to the flexibility of the system and its resistance to central shocks.

There have been not in my back yard" (NIMBY) concerns relating to the visual and other impacts of some wind farms, with local residents sometimes fighting or blocking construction. In the USA, the Massachusetts Cape Wind project was delayed for years partly because of aesthetic concerns. However, residents in other areas have been more positive. According to a town councilor, the overwhelming majority of locals believe that the Ardrossan Wind Farm in Scotland has enhanced the area.

A recent UK Government document states that projects are generally more likely to succeed if they have broad public support and the consent of local communities. This means giving communities both a say and a stake". In countries such as Germany and Denmark many renewable projects are owned by communities, particularly through cooperative structures, and contribute significantly to overall levels of renewable energy deployment.

The market for renewable energy technologies has continued to grow. Climate change concerns and increasing in green jobs, coupled with high oil prices, peak oil, oil wars, oil spills, promotion of electric vehicles and renewable electricity, nuclear disasters and increasing government support, are driving increasing renewable energy legislation, incentives and commercialization. New government spending, regulation and policies helped the industry weather the 2009 economic crisis better than many other sectors.

REFERENCES

1. Ipsos 2011, p. 3
2. Ellabban, Omar; Abu-Rub, Haitham; Blaabjerg, Frede (2014). Renewable energy resources: Current status, future prospects and their enabling technology". Renewable and Sustainable Energy Reviews. **39**: 748–764 [749]. doi:10.1016/j.rser.2014.07.113.
3. REN21 (2010). Renewables 2010 Global Status Report p. 15.
4. REN21, Global Status Report 2016. Retrieved 8th June 2016.
5. IRENA, Renewable energy and jobs, Annual review 2015, IRENA.
6. https://www.theguardian.com/environment/2016/oct/25/renewables-made-up-half-of-net-electricity-capacity-added-last-year
7. International Energy Agency (2012). Energy Technology Perspectives 2012" (PDF).
8. Heidari, Negin; Pearce, Joshua M. (2016). A Review of Greenhouse Gas Emission Liabilities as the Value of Renewable Energy for Mitigating Lawsuits for Climate Change Related Damages". Renewable and Sustainable Energy Reviews. **55C**: 899–908. doi:10.1016/j.rser.2015.11.025.
9. "Global Trends in Sustainable Energy Investment 2007: Analysis of Trends and Issues in the Financing of Renewable Energy and Energy Efficiency in OECD and Developing Countries" (PDF). unep.org. United Nations Environment Programme. 2007. p. 3. Archived from the original on 13 October 2014. Retrieved 13 October 2014.
10. REN21 (2013). "Renewables global futures report 2013" (PDF).
11. Vad Mathiesen, Brian; et al. (2015). Smart Energy Systems for coherent 100% renewable energy and transport solutions". Applied Energy. **145**: 139–154. doi:10.1016/j.apenergy.2015.01.075.
12. World Energy Assessment (2001). Renewable energy technologies, p. 221.
13. Steve Leone (25 August 2011). U.N. Secretary-General: Renewables Can End Energy Poverty". Renewable Energy World.
14. Armaroli, Nicola; Balzani, Vincenzo (2011). Towards an electricity-powered world". Energy and Environmental Science. **4**: 3193–3222. doi:10.1039/c1ee01249e.
15. Armaroli, Nicola; Balzani, Vincenzo (2016). Solar Electricity and Solar Fuels: Status and Perspectives in the Context of the Energy Transition". Chemistry – A

European Journal. **22**: 32–57. doi:10.1002/chem.201503580.

16. Volker Quaschning, Regenerative Energiesysteme. Technologie – Berechnung – Simulation. 8th. Edition. Hanser (Munich) 2013, p. 49.
17. IEA Renewable Energy Working Party (2002). Renewable Energy... into the mainstream, p. 9.
18. Jacobson, Mark Z.; et al. (2015). : 100% clean and renewable wind, water, and sunlight (WWS) all-sector energy roadmaps for the 50 United States". Energy and Environmental Science. **8**: 2093–2117. doi:10.1039/C5EE01283J.
19. Schröder, K.-P.; Smith, R.C. (2008). Distant future of the Sun and Earth revisited". Monthly Notices of the Royal Astronomical Society. **386** *(1): 155–163.* arXiv:0801.4031. Bibcode:2008MNRAS.386..155S. doi:10.1111/j.1365-2966.2008.13022.x. See also Palmer, J. (2008). Hope dims that Earth will survive Sun's death". New Scientist. Retrieved 24 March 2008.
20. Carrington, D. (21 February 2000). Date set for desert Earth". BBC News. Retrieved 31 March 2007.
21. Clean Edge (2009). Clean Energy Trends 2009 pp. 1–4.
22. Ben Sills (29 August 2011). Solar May Produce Most of World's Power by 2060, IEA Says". Bloomberg.
23. REN21 (2011). Renewables 2011: Global Status Report" (PDF). p. 14.
24. REN21. Renewables 2016: Global Status Report" (PDF). Renewable Energy Policy Network. Retrieved 2016-10-31.
25. REN21 (2010). Renewables 2010 Global Status Reportp. 53.
26. K. Kris Hirst. The Discovery of Fire". About.com. Retrieved 15 January 2013.
27. wind energy". The Encyclopedia of Alternative Energy and Sustainable Living. Retrieved 15 January 2013.
28. The surprising history of sustainable energy". Sustainablehistory.wordpress.com. Archived from the original on Dec 24, 2014. Retrieved 1 November 2012.
29. Werner von Siemens (1885). On the electromotive action of illuminated selenium, discovered by Mr. Fritts, of New York". Van Nostrands Engineering Magazine. **32**: 514–516.
30. Weber suggests that the modern economic world will determine the lifestyle of everyone born into it "until the last hundredweight of fossil fuel is burned" (bis der letzte Zentner fossilen Brennstoffs verglüht ist).
31. "Power from Sunshine": A Business History of Solar Energy 25 May 2012
32. Hubbert, M. King (June 1956). Nuclear Energy and the Fossil Fuels" (PDF). Shell Oil Company/American Petroleum Institute. Retrieved 10 November 2014.
33. History of PV Solar". Solarstartechnologies.com. Archived from the original on 6 December 2013. Retrieved 1 November 2012.
34. Tweed, Katherine. In 2040, Fossil Fuels Still Reign". IEEE. Retrieved 15 November 2014.

35. Analysis of Wind Energy in the EU-25" (PDF). European Wind Energy Association. Retrieved 11 March 2007.
36. Martin Kaltschmitt, Wolfgang Streicher, Andreas Wiese (eds.): Erneuerbare Energien. Systemtechnik, Wirtschaftlichkeit, Umweltaspekte. Springer, Berlin/ Heidelberg 2013, p. 819.
37. "Offshore stations experience mean wind speeds at 80 m that are 90% greater than over land on average. Evaluation of global wind power
38. Overall, the researchers calculated winds at 80 meters [300 feet] above sea level traveled over the ocean at approximately 8.6 meters per second and at nearly 4.5 meters per second over land [20 and 10 miles per hour, respectively]." Global Wind Map Shows Best Wind Farm Locations . Retrieved 30 January 2006.
39. http://www.bp.com/content/dam/bp/excel/energy-economics/statistical-review-2015/bp-statistical-review-of-world-energy-2015-workbook.xlsx
40. http://www.ren21.net/wp-content/uploads/2016/06/GSR_2016_Full_Report_REN21.pdf
41. https://www.irena.org/DocumentDownloads/Publications/RE_Technologies_Cost_Analysis-HYDROPOWER.pdf
42. Worldwatch Institute (January 2012). Use and Capacity of Global Hydropower Increases".
43. ^ Jump up to:[a] [b] Solar Energy Perspectives: Executive Summary". International Energy Agency. 2011. Archived from the original (PDF) on 3 December 2011.
44. Solar Fuels and Artificial Photosynthesis". Royal Society of Chemistry. 2012. Retrieved 11 March 2013.
45. Energy Sources: Solar". Department of Energy. Retrieved 19 April 2011.
46. NREL.gov U.S. Renewable Energy Technical Potentials: A GIS-Based Analysis, July 2013
47. thinkprogress.org National Renewable Energy Laboratory: Solar Has The Most Potential Of Any Renewable Energy Source, 30 July 2013
48. http://www.ren21.net/wp-content/uploads/2016/10/REN21_GSR2016_FullReport_en_11.pdf pg64
49. Dye, S. T. (2012). Geoneutrinos and the radioactive power of the Earth". Reviews of Geophysics. **50**: 3. Bibcode:2012RvGeo..50.3007D. doi:10.1029/2012rg000400.
50. Gando, A.; Dwyer, D. A.; McKeown, R. D.; Zhang, C. (2011). Partial radiogenic heat model for Earth revealed by geoneutrino measurements". Nature Geoscience. **4** *(9): 647–651.* doi:10.1038/ngeo1205.
51. Nemzer, J. Geothermal heating and cooling".
52. [3], DSIRE: Database of State Incentives for Renewables & Efficiency
53. Zero-energy building
54. Net Zero Foundation, Net Zero Technologies
55. New York City Council Gives a Thumbs Up to Geothermal Heat Pumps,

December 14, 2015

56. U.S. EPA, The Social Cost of Carbon
57. Sci Am, What Is the Right Price for Carbon? Evan Lehmann, Climatewire on April 2, 2010
58. Biomass Energy Center. Biomassenergycentre.org.uk. Retrieved on 28 February 2012.
59. Retrieved on 12 April 2012.
60. T.A. Volk, L.P. Abrahamson (January 2000). Developing a Willow Biomass Crop Enterprise for Bioenergy and Bioproducts in the United States". North East Regional Biomass Program. Retrieved 4 June 2015.
61. Energy crops". crops are grown specifically for use as fuel. BIOMASS Energy Centre. Retrieved 6 April 2013.
62. Energy Kids. Eia.doe.gov. Retrieved on 28 February 2012.
63. Fuel Ethanol Production: GSP Systems Biology Research". U.S. Department of Energy Office of Science. 19 April 2010. Archived from the original on 27 May 2010. Retrieved 2 August 2010.
64. Breaking the Biological Barriers to Cellulosic Ethanol: A Joint Research Agenda" (PDF). June 2006. Retrieved 2 August 2010.
65. Frauke Urban and Tom Mitchell 2011. Climate change, disasters and electricity generation. London: Overseas Development Institute and Institute of Development Studies
66. Demirbas, A. . (2009). Political, economic and environmental impacts of biofuels: A review". Applied Energy. **86**: S108–S117. doi:10.1016/j.apenergy.2009.04.036.
67. Sweet sorghum for food, feed and fuel New Agriculturalist, January 2008.
68. Opinion of the EEA Scientific Committee on Greenhouse Gas Accounting in Relation to Bioenergy". Retrieved 1 November 2012.
69. REN21 (2011). Renewables 2011: Global Status Report" (PDF). pp. 13–14. Archived from the original(PDF) on 13 May 2012.
70. WHO - 7 million premature deaths annually linked to air pollution".
71. WHO - Ambient (outdoor) air quality and health".
72. WHO - Household air pollution and health".
73. REN21, Renewables Global Status Report 2012". Ren21.net. Archived from the original (PDF) on 11 August 2014. Retrieved 11 August 2014.
74. GWEC Global Wind Statistics 2014" (PDF). GWEC. 10 February 2015.
75. The World Wind Energy Association (2014). 2014 Half-year Report. WWEA. pp. 1–8.
76. Wind in power: 2015 European statistics- EWEA
77. Tam Hunt (9 March 2015). The Solar Singularity Is Nigh". Greentech Media. Retrieved 29 April 2015.
78. World largest solar thermal plant syncs to the grid". Spectrum.ieee.org. Retrieved

28 November 2014.

79. "World's Largest Solar Thermal Power Project at Ivanpah Achieves Commercial Operation", NRG press release, February 13, 2014.

80. http://www.ren21.net/wp-content/uploads/2015/06/GSR2015_Figure25.jpg

81. REN21 (2011). Renewables 2011: Global Status Report" (PDF). p. 15.

82. REN21 (2012). Renewables Global Status Report 2012p. 17.

83. REN21 2013 Renewables Global Status Report" (PDF). Retrieved 30 January 2014.

84. REN21. Renewables 2014: Global Status Report" (PDF). Archived from the original on 4 September 2014. Retrieved 20 January 2015.

85. http://www.ren21.net/wp-content/uploads/2015/07/REN12-GSR2015_Onlinebook_low1.pdf

86. E. Lantz, M. Hand, and R. Wiser (13–17 May 2012) "The Past and Future Cost of Wind Energy," National Renewable Energy Laboratory conference paper no. 6A20-54526, page 4

87. Solar Photovoltaics Competing in the Energy Sector—On the road to competitiveness" (PDF). European Photovoltaic Industry Association. September 2011. p. 18. Retrieved March 2015. Note: Germany already reached cost-range of €0.08–0.14/kWh in 2013. Check date values in: |access-date=(help)

88. Henning Gloystein (23 November 2011). Renewable energy becoming cost competitive, IEA says". Reuters.

89. International Renewable Energy Agency (2012). Renewable Power Generation Costs in 2012: An Overview" (PDF).

90. Timmer, John (25 September 2013). Cost of renewable energy's variability is dwarfed by the savings: Wear and tear on equipment costs millions, but fuel savings are worth billions.". Ars Technica. Condé Nast. Retrieved 26 September 2013.

91. How much hydro power could be built at undeveloped sites?".

92. GWEC Global Wind Statistics 2014" (PDF). GWEC. 10 February 2015.

93. World Wind Energy Report 2010" (PDF). Report. World Wind Energy Association. February 2011. Retrieved 30 April 2011.

94. Renewables". eirgrid.com. Retrieved 22 November 2010.

95. Terra-Gen Closes on Financing for Phases VII and IX, Business Wire, 17 April 2012

96. Patrick Barkham (8 January 2009). Blown away". London: Guardian. Retrieved 21 November 2011.

97. Segs Iii, Iv, V, Vi, Vii, Viii & Ix". Fplenergy.com. Retrieved 31 January 2012.

98. Brightsource Ivanpah". ivanpahsolar.com. Retrieved 16 May 2014.

99. Mearian, Lucas. U.S. flips switch on massive solar power array that also stores electricity: The array is first large U.S. solar plant with a thermal energy storage

system, 10 October 2013. Retrieved 18 October 2013.

100. Global Concentrating Solar Power" (PDF). International Renewable Energy Agency. June 2012. Retrieved 8 September 2012.
101. Solar Thermal Projects Under Review or Announced". Energy.ca.gov. Retrieved 21 November 2011.
102. REN21 (2008). Renewables 2007 Global Status Report (PDF) p. 12.
103. Crossing the Chasm" (PDF). Deutsche Bank Markets Research. 27 February 2015. Archived from the original on 1 April 2015.
104. 2014 Outlook: Let the Second Gold Rush Begin" (PDF). Deutsche Bank Markets Research. 6 January 2014. Archived from the original on 22 November 2014. Retrieved 22 November 2014.
105. GreenTechMedia.com, RenewEconomy, Giles Parkinson Deutsche Bank Predicts Second Solar ‹Gold Rush', 9 January 2014
106. Giles Parkinson (13 August 2014). Citigroup: Outlook for global solar is getting brighter". RenewEconomy. Retrieved 18 August 2014.
107. Snapshot of Global PV 1992-2014" (PDF). iea-pvps.org. International Energy Agency — Photovoltaic Power Systems Programme. 30 March 2015. Archived from the original on 30 March 2015.
108. Global Market Outlook for Photovoltaics 2014-2018"(PDF). epia.org. EPIA – European Photovoltaic Industry Association. Archived from the original on 12 June 2014. Retrieved 12 June 2014.
109. iea.org (2014). Technology Roadmap: Solar Photovoltaic Energy" (PDF). IEA. Archived from the original on 7 October 2014. Retrieved 7 October 2014.
110. Denis Lenardic. Large-scale photovoltaic power plants ranking 1 - 50 PVresources. com, 2010.
111. Solar Integrated in New Jersey". Jcwinnie.biz. Retrieved 20 August 2013.
112. IEA says biofuels can displace 27% of transportation fuels by 2050 Washington". Platts. 20 April 2011.
113. Industry Statistics: Annual World Ethanol Production by Country". Renewable Fuels Association. Archived from the original on 8 April 2008. Retrieved 2 May 2008.
114. Macedo Isaias, M. Lima Verde Leal and J. Azevedo Ramos da Silva (2004). Assessment of greenhouse gas emissions in the production and use of fuel ethanol in Brazil" (PDF). Secretariat of the Environment, Government of the State of São Paulo. Archived from the original (PDF) on 28 May 2008. Retrieved 9 May 2008.
115. Daniel Budny and Paulo Sotero, editor (April 2007). Brazil Institute Special Report: The Global Dynamics of Biofuels"(PDF). Brazil Institute of the Woodrow Wilson Center. Retrieved 3 May 2008.
116. Erica Gies. As Ethanol Booms, Critics Warn of Environmental Effect The New York Times, 24 June 2010.

117. American Energy: The Renewable Path to Energy Security" (PDF). Worldwatch Institute. September 2006. Retrieved 11 March 2007.
118. William E. Glassley. Geothermal Energy: Renewable Energy and the Environment CRC Press, 2010.
119. Geothermal Energy Association. Geothermal Energy: International Market Update May 2010, p. 4-6.
120. Geothermal Energy Association. Geothermal Energy: International Market Update - May 2010, p. 7
121. Khan, M. Ali (2007). The Geysers Geothermal Field, an Injection Success Story" (PDF). Annual Forum of the Groundwater Protection Council. Archived from the original(PDF) on 26 July 2011. Retrieved 25 January 2010.
122. FS-UNEP Collaborating Centre, Global trends in renewable energy investment, 2016 p. 14.
123. Power for the People p. 3. Archived 30 March 2012 at the Wayback Machine.
124. Kevin Bullis (27 January 2012). In the Developing World, Solar Is Cheaper than Fossil Fuels". Technology Review.
125. REN21 (2010). Renewables 2010 Global Status Reportp. 12. Archived 13 May 2012 at the Wayback Machine.
126. Fry, Carolyn. 28 June 2012. Anguilla moves towards cleaner energy
127. Ethiopia". Projectgaia.com. Retrieved 1 November 2012.
128. Energy for Development: The Potential Role of Renewable Energy in Meeting the Millennium Development Goals pp. 7-9.
129. Bloomberg New Energy Finance, UNEP SEFI, Frankfurt School, Global Trends in Renewable Energy Investment 2011". Unep.org. Retrieved 21 November 2011.
130. REN21 (2008). Renewables 2007 Global Status Report (PDF) p. 18.
131. Ken Berlin, Reed Hundt, Marko Muro, and Devashree Saha. "State Clean Energy Banks: New Investment Facilities for Clean Energy Deployment"
132. Hooper, Craig (2011). Air Force cedes the Green lead–and the lede–to Navy". nextnavy.com. Retrieved 27 December 2011.
133. Signatory States Archived 26 December 2010 at the Wayback Machine.
134. Signatories of IRENA's statute Archived 24 January 2011 at the Wayback Machine.
135. Mark Tran (2 November 2011). UN calls for universal access to renewable energy". The Guardian. London.
136. Paul Gipe (4 April 2013). 100 Percent Renewable Vision Building". Renewable Energy World.
137. S. Pacala & R. Socolow (2004). Stabilization Wedges: Solving the Climate Problem for the Next 50 Years with Current Technologies" (PDF). Science Vol. 305. pp. 968–972.
138. Sørensen, Bent (1975). A plan is outlined according to which solar and wind

energy would supply Denmark's needs by the year 2050". Science. **189** *(4199): 255–260. doi:10.1126/science.189.4199.255.*

139. Lund, Henrik (2006). Large-scale integration of optimal combinations of PV, wind and wave power into the electricity supply". Renewable Energy. **31** *(4): 503–515. doi:10.1016/j.renene.2005.04.008.*

140. Hohmeyer, Olav; Bohm, Sönke (2015). Trends toward 100% renewable electricity supply in Germany and Europe: a paradigm shift in energy policies". Wiley Interdisciplinary Reviews: Energy and Environment. **4**: 74–97. doi:10.1002/wene.128.

141. Mark A. Delucchi & Mark Z. Jacobson (2011). Providing all global energy with wind, water, and solar power, Part II: Reliability, system and transmission costs, and policies"(PDF). Energy Policy. Elsevier Ltd. pp. 1170–1190.

142. National Research Council (2010). Electricity from Renewable Resources: Status, Prospects, and Impediments". National Academies of Science. p. 4.

143. John Wiseman; et al. (April 2013). Post Carbon Pathways" (PDF). University of Melbourne.

144. International Energy Agency (2007). Renewables in global energy supply: An IEA facts sheet (PDF), OECD, p. 3. Archived 12 October 2009 at the Wayback Machine.

145. S.C.E. Jupe; A. Michiorri; P.C. Taylor (2007). Increasing the energy yield of generation from new and renewable energy sources". Renewable energy. **14** *(2): 37–62.*

146. Defense-scale supercomputing comes to renewable energy research". Sandia National Laboratories. Retrieved 16 April 2012.

147. Sandia National Laboratories" (PDF). Sandia National Laboratories. Retrieved 16 April 2012.

148. *Chakrabarty, Gargi, 16 April 2009. Stimulus leaves NREL in cold" Denver Post"

149. Duchane, Dave; Brown, Don (December 2002). Hot Dry Rock (HDR) Geothermal Energy Research and Development at Fenton Hill, New Mexico" (PDF). Geo-Heat Centre Quarterly Bulletin. **23** *(4). Klamath Falls, Oregon: Oregon Institute of Technology. pp. 13–19. ISSN 0276-1084. Retrieved 5 May 2009.*

150. Australia›s Renewable Energy Future inc Cooper Basin & geothermal map of Australia Retrieved 15 August 2015

151. Dyadic International - Bioenergy , Biopharmaceutical Enzymes".

152. Pernick, Ron and Wilder, Clint (2007). The Clean Tech Revolution p. 96.

153. Carbon Trust, Future Marine Energy. Results of the Marine Energy Challenge: Cost competitiveness and growth of wave and tidal stream energy, January 2006

154. Sihwa Tidal Power Plant". Renewable Energy News and Articles.

155. Tidal power (PDF), retrieved 20 March 2010

156. Kyocera, partners announce construction of the world's largest floating solar

PV Plant in Hyogo prefecture, Japan". SolarServer.com. 4 September 2014.

157. Running Out of Precious Land? Floating Solar PV Systems May Be a Solution". EnergyWorld.com. 7 November 2013.
158. Vikram Solar commissions India's first floating PV plant". SolarServer.com. 13 January 2015.
159. Sunflower Floating Solar Power Plant In Korea". CleanTechnica. 21 December 2014.
160. Short Of Land, Singapore Opts For Floating Solar Power Systems". CleanTechnica. 5 May 2014.
161. Erica Goodemay, New Solar Plants Generate Floating Green Power, New York Times, 20 May 2016.
162. Winery goes solar with Floatovoltaics". SFGate. 29 May 2008. Retrieved 31 May 2013.
163. Yamakura Dam in Chiba Prefecture". The Japan Dam Foundation. Retrieved 1 February 2015.
164. Kyocera and Century Tokyo Leasing to Develop 13.4MW Floating Solar Power Plant on Reservoir in Chiba Prefecture, Japan, Kyocera, December 22, 2014
165. New Solar Plants Generate Floating Green Power NYT May 20, 2016
166. Solar Panels Floating on Water Could Power Japan's Homes, National Geographic, Bryan Lufkin, January 16, 2015
167. Anand Upadhyay (6 April 2015). Brazil Announces Huge 350 MW Floating Solar Power Plant". CleanTechnica.com.
168. Solar-assisted heat pumps". Retrieved 21 June 2016.
169. Pompe di calore elio-assistite" (in Italian). Retrieved 21 June 2016.
170. Collings AF and Critchley C (eds). Artificial Photosynthesis – From Basic Biology to Industrial Application (Wiley-VCH Weinheim 2005) p ix.
171. Energy and environment policy case for a global project on artificial photosynthesis". Energy & Environmental Science. RSC Publishing. **6**: 695. doi:10.1039/C3EE00063J. Retrieved 19 August 2013.
172. jobs. 'Artificial leaf' faces economic hurdle: Nature News & Comment". Nature.com. Retrieved 7 November 2012.
173. Power Beaming
174. Noth, André (July 2008). History of Solar Flight" (PDF). Autonomous Systems Lab. Zürich: Swiss Federal Institute of Technology. p. 3. Retrieved 8 July 2010. Günter Rochelt was the designer and builder of Solair I, a 16 m wingspan solar airplane ... 21st of August 1983 he flew in Solair I, mostly on solar energy and also thermals, during 5 hours 41 minutes.
175. Infographic: A Timeline Of The Present And Future Of Electric Flight". Popular Science. Retrieved 7 January 2016.
176. Taylor, John W R (1974). Jane's All the World's Aircraft 1974-75. London:

Jane's Yearbooks. p. 573. ISBN 0 354 00502 2.

177. Batrawy, Aya (9 March 2015). Solar-powered plane takes off for flight around the world". Associated Press. Retrieved 14 March 2015.
178. http://spacejournal.ohio.edu/issue18/thermalpower.html
179. Leighty and Holbrook (2012) Running the World on Renewables: Alternatives for Transmission and Low-cost Firming Storage of Stranded Renewables as Hydrogen and Ammonia Fuels via Underground Pipelines" Proceedings of the ASME 2012 International Mechanical Engineering Congress & Exposition 9–15 November 2012, Houston, Texas
180. Graves, Christopher; Ebbesen, Sune D.; Mogensen, Mogens; Lackner, Klaus S. (2011). Sustainable hydrocarbon fuels by recycling CO_2 *and* H_2O with renewable or nuclear energy". Renewable and Sustainable Energy Reviews. **15** *(1): 1–23. doi:10.1016/j.rser.2010.07.014.* (Review.)
181. Lackner, Klaus S.; et al. (2012). The urgency of the development of CO_2 *capture from ambient air". Proceedings of the National Academy of Sciences of the United States of America.* **109** *(33): 13156–62. Bibcode*:2012PNAS..10913156L. doi:10.1073/pnas.1108765109. PMID 22843674. Retrieved 7 September 2012.
182. Pearson, R.J.; Eisaman, M.D.; et al. (2012). Energy Storage via Carbon-Neutral Fuels Made From CO_2, Water, and Renewable Energy" (PDF). Proceedings of the IEEE. **100** *(2): 440–60. doi:10.1109/JPROC.2011.2168369. Archived from the original (PDF) on 12 May 2013. Retrieved 7 September 2012.* (Review.)
183. Center for Solar Energy & Hydrogen Research Baden-Württemberg (2011). Verbundprojekt ‹Power-to-Gas›" (in German). zsw-bw.de. Retrieved 9 September 2012.
184. Center for Solar Energy & Hydrogen Research (24 July 2012). Bundesumweltminister Altmaier und Ministerpräsident Kretschmann zeigen sich beeindruckt von Power-to-Gas-Anlage des ZSW" (in German). zsw-bw.de. Retrieved 9 September 2012.
185. Fraunhofer-Gesellschaft (5 May 2010). Storing green electricity as natural gas". fraunhofer.de. Retrieved 9 September 2012.
186. "George Olah CO2 to Renewable Methanol Plant, Reykjanes, Iceland" (Chemicals-Technology.com)
187. "First Commercial Plant" (Carbon Recycling International)
188. International Energy Agency (2007). Contribution of Renewables to Energy Security IEA Information Paper, p. 5. Archived 18 March 2009 at the Wayback Machine.
189. Whatever Happened to Wind Energy?". LiveScience. 14 January 2008. Retrieved 17 January 2012.
190. Simon Gourlay (12 August 2008). Wind farms are not only beautiful, they're absolutely necessary". The Guardian. UK. Retrieved 17 January 2012.
191. Department of Energy & Climate Change (2011). UK Renewable Energy Roadmap (PDF) p. 35.

192. DTI, Co-operative Energy: Lessons from Denmark and Sweden, Report of a DTI Global Watch Mission, October 2004
193. Morris C & Pehnt M, German Energy Transition: Arguments for a Renewable Energy Future, Heinrich Böll Foundation, November 2012
194. Spellman, Frank R. (2013). Safe Work Practices for Green Energy Jobs (first ed.). DEStech Publications. p. 323. ISBN 978-1-60595-075-4. Retrieved 29 December 2014.

CHAPTER 6

Rooftop Farming

Chapter Contents

INTRODUCTION

A roof garden is a garden on the roof of a building. Besides the decorative benefit, roof plantings may provide food, temperature control, hydrological benefits, architectural enhancement, habitats or corridors for wildlife, recreational opportunities, and in large scale it may even have ecological benefits. The practice of cultivating food on the rooftop of buildings is sometimes referred to as rooftop farming. Rooftop farming is usually done using green roof, hydroponics, aeroponics or air-dynaponics systems or container gardens.

Roof garden of Rockefeller Center in Manhattan

Sky garden at 20 Fenchurch Street

The roof terrace of the Casa Grande hotel in Santiago de Cuba.

Humans have grown plants atop structures since antiquity. The ziggurats of ancient Mesopotamia (4th millennium BC–600 BC) had plantings of trees and shrubs on aboveground terraces. An example in Roman times was the Villa of the Mysteries in Pompeii, which had an elevated terrace where plants were grown. A roof garden has also been discovered around an audience hall in Roman-Byzantine Caesarea. The medieval Egyptian city of Fustat had a number of high-rise buildings that Nasir Khusraw in the early 11th century described as rising up to 14 stories, with roof gardens on the top story complete with ox-drawn water wheels for irrigating them.

Among the Seven Wonders of the Ancient World, The Hanging Gardens are often depicted as tall structures holding vegetation; even immense trees.

ENVIRONMENTAL IMPACT

View of ACROS Fukuoka designed by architect Emilio Ambasz.

Infinity edge pool at Sands Sky Park, Marina Bay Sands Hotel, Singapore

Roof gardens are most often found in urban environments. Plants have the ability to reduce the overall heat absorption of the building which then reduces energy consumption. "The primary cause of heat build-up in cities is insolation, the absorption of solar radiation by roads and buildings in the city and the storage of this heat in the building material and its subsequent re-radiation. Plant surfaces however, as a result of transpiration, do not rise more than 4–5 °C above the ambient and are sometimes cooler." This then translates into a cooling of the environment between 3.6 and 11.3 degrees Celsius (6.5 and 20.3 °F), depending on the area on earth (in hotter areas, the environmental temperature will cool more). The study was performed by the University of Cardiff.

A study at the National Research Council of Canada showed the differences between roofs with gardens and roofs without gardens against temperature. The study shows temperature effects on different layers of each roof at different times of the day. Roof gardens are obviously very beneficial in reducing the effects of temperature against roofs without gardens. "If widely adopted, rooftop gardens could reduce the urban heat island, which would decrease smog episodes, problems associated with heat stress and further lower energy consumption."

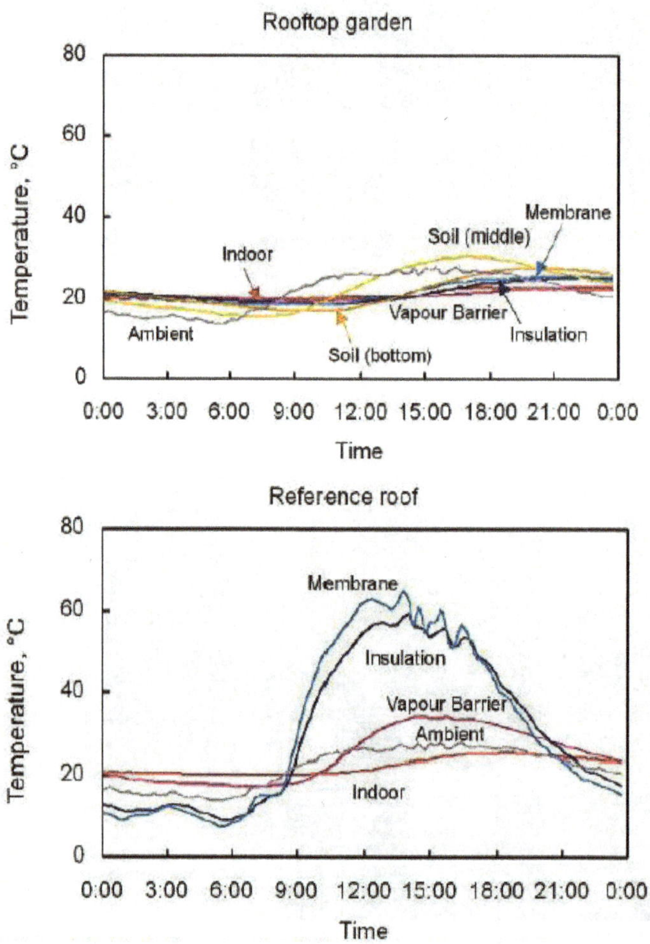

Temperature profiles of rooftop gardens vs reference roof at different times of a certain day

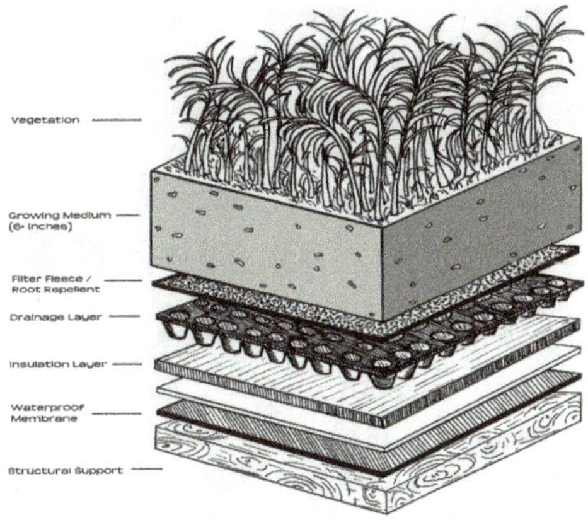

Aside from rooftop gardens providing resistance to thermal radiation, rooftop gardens are also beneficial in reducing rain run off. A roof garden can delay run off; reduce the rate and volume of run off. "As cities grow, permeable substrates are replaced by impervious structures such as buildings and paved roads. Storm water run-off and combined sewage overflow events are now major problems for many cities in North America. A key solution is to reduce peak flow by delaying (e.g., control flow drain on roofs) or retaining run-off (e.g., rain detention basins). Rooftop gardens can delay peak flow and retain the run-off for later use by the plants."

Urban Agriculture

Palais de Tokyo, Paris

"In an accessible rooftop garden, space becomes available for localized small-scale urban agriculture, a source of local food production. An urban garden can supplement the diets of the community it feeds with fresh produce and provide a tangible tie to food production." At Trent University, there is currently a working rooftop garden which provides food to the student café and local citizens.

Available gardening areas in cities are often seriously lacking, which is likely the key impetus for many roof gardens. The garden may be on the roof of an autonomous building which takes care of its own water and waste. Hydroponics and other alternative methods can expand the possibilities of roof top gardening by reducing, for example, the need for soil or its tremendous weight. Plantings in containers are used extensively in roof top gardens. Planting in containers prevents added stress to the roof's waterproofing. One high-profile example of a building with a roof garden is Chicago City Hall.

For those who live in small apartments with little space, square foot gardening, or (when even less space is available) green walls (vertical gardening) can be a solution. These use much less space than traditional gardening (square foot gardening uses 20% of the space of conventional rows; ten times more produce can be generated from vertical gardens). These also encourage environmentally responsible practices, eliminating tilling, reducing or eliminating pesticides, and weeding, and encouraging the recycling of wastes through composting.

Importance to Urban Planning

Hotel Astor roof garden, ca. 1904

Becoming green is a high priority for urban planners. The environmental and aesthetic benefits to cities is the prime motivation. It was calculated that the temperature in Tokyo could be lowered by 0.11–0.84 °C if 50% of all available rooftop space were planted with greenery. This would lead to a savings of approximately 100 million yen

Singapore is active in green urban development. "Roof gardens present possibilities for carrying the notions of nature and open space further in tall building development."

When surveyed, 80% of Singapore residents voted for more roof gardens to be implemented in the city's plans. Recreational reasons, such as leisure and relaxation, beautifying the environment, and greenery and nature, received the most amount of votes. Planting roof gardens on tops of building is a way to make cities more efficient.

Roof Garden vs. Green Roof

A roof garden can be distinguished from a green roof, although the two terms are often used interchangeably. The term roof garden is well suited to roof spaces that incorporate recreation, entertaining, and provide additional outdoor living space for the building's residents. It may include planters, plants, dining and lounging furniture, outdoor structures such as pergolas and sheds, and automated irrigation and lighting systems.

Although they may provide aesthetic and recreational benefits a green roof is not necessarily designed for this purpose. A green roof may not provide any recreational space and be constructed with an emphasis towards improving the insulation or improving the overall energy efficiency and reducing the cooling and heating costs within a building.

Green roofs may be extensive or intensive. The terms are used to describe the type of planting required. The panels that comprise a green roof are generally no more than a few inches up to a foot in depth, since weight is an important factor when covering an entire roof surface. The plants that go into a green roof are usually sedum or other shallow-rooted plants that will tolerate the hot, dry, windy conditions that prevail on most rooftop gardens. With a green roof, "the plants layer can shield off as much as 87% of solar radiation while a bare roof receives 100% direct exposure".

The planters on a roof garden may be designed for a variety of functions and vary greatly in depth to satisfy aesthetic and recreational purposes. These planters can hold a range of ornamental plants: anything from trees, shrubs, vines, or an assortment of flowers. As aesthetics and recreation are the priority they may not provide the environmental and energy benefits of a green roof.Planting on roof tops can make urban living more self-sufficient and make fresh vegetables more accessible to urban people.

Science of Gardening

The related idea of a living machine is based on the most basic cycle of gardening: using wastes (organic waste and sewage), appropriately broken down, usually in some specialized container, on the soil, and harvesting food which, when processed, generates biodegradable waste, and when eaten, generates sewage. In most of the world, this kind of very tight closed loop gardening is used, despite certain health risks if necessary precautions are not taken. Composting human or pet waste should achieve thermophilic conditions and age for at least a year before being used.

In Popular Culture

- American jazz singer Al Jarreau composed a song named "Roof Garden".

- Apu Nahasapeemapetilon of the TV show *The Simpsons* has a roof top garden visited by Paul McCartney and his wife.
- In BBC's 1990 television miniseries *House of Cards*, the main character, MP Francis Urquhart, murders journalist Mattie Storin by throwing her off of the Palace of Westminster's rooftop garden.

Gallery

Sveavägen 44 uses Sedum

Sveavägen 44

Villa Arson

Villa Arson

ACROS Fukuoka

Herb garden of National Museum of Nature and Science Tokyo

A small Zen roof garden

Inari Daimyojin on the roof of Hankyu Department Stores Umeda main store

Tokyo Plaza

Canal City Hakata

REFERENCES

1. Louise Lundberg Scandinavian Green Roof Institute (2009). "The benefits of Rooftop Gardens" (PDF). Retrieved March 12, 2014.
2. "World's Largest Rooftop Farm Documents Incredible Growth High Above Brooklyn". The Huffington Post. February 12, 2014. Retrieved March 12, 2014.
3. Michelle Nowak (May 2004). "Urban Agriculture on the Rooftop". City Farmer, Canada's Office of Urban Agriculture. Retrieved March 12, 2014.
4. pp. 112–115, chapter 2, "Roof gardens through history", Roof gardens: history, design, and construction, Theodore Osmundson, W. W. Norton & Company, 1999, ISBN 0-393-73012-3.
5. p. 219, Byzantine garden culture, Antony Robert Littlewood, Henry Maguire, and Joachim Wolschke-Bulmahn, Dumbarton Oaks, 2002, ISBN 0-88402-280-3.
6. Behrens-Abouseif, Doris (1992), Islamic Architecture in Cairo, Brill Publishers, p. 6, ISBN 90-04-09626-4
7. Ong, B. (2003). Green plot ratio: an ecological measure for architecture and urban planning. Landscape and Urban Planning, 63 (4). Retrieved June 19, 2009, from ScienceDirect database.
8. Catherine Brahic (September 28, 2007). "Cooling percentages by 'Green roofs'". NewScientist. Retrieved March 12, 2014.
9. Liu, K. "Energy efficiency and environmental benefits of rooftop gardens" (PDF). National Research Council Canada. Retrieved March 12, 2014.
10. Trent University (October 30, 2009). "Rooftop Gardens" (PDF). Retrieved March 12, 2014.
11. Yuen, B, & Wong, N. (2005). Resident perceptions and expectations of rooftop gardens in Singapore. Landscape and Urban Planning, 73 (4). Retrieved June 19, 2009, from ScienceDirect database.
12. Zimmerman, Greg. "Extensive or Intensive? Green Roofs Explained". Retrieved 17 January 2017.
13. Wong, N, Tay, S, Wong, R, Ong, C, & Sia, A. (2003). Life cycle cost analysis of rooftop gardens in Singapore. Building and Environment, 38 (3). Retrieved June 19, 2009, from ScienceDirect database.

CHAPTER 7

Water Chiller

Chapter Contents

- **Introduction**
- **Water Cooling**
- **Nomenclature**
- **Advantages**
- **Open Method**
- **Environmental Considerations**
- **Cooling Water Chemistry**

INTRODUCTION

A water chiller is a mechanical device used to facilitate heat exchange from water to a refrigerant in a closed loop system. The refrigerant is then pumped to a location where the waste heat is transferred to the atmosphere.

In hydroponics, pumps, lights and ambient heat can warm the reservoir water temperatures, leading to plant root and health problems. For ideal plant health, a chiller can be used to lower the water temperature below ambient level; 68°F (20°C) is a good temperature for most plants. This results in healthy root production and efficient absorption of nutrients.

In air conditioning, chilled water is often used to cool a building's air and equipment, especially in situations where many individual rooms must be controlled separately, such as a hotel. A chiller lowers water temperature to between 40° and 45°F before the water is pumped to the location to be cooled.

WATER COOLING

Water cooling is a method of heat removal from components and industrial equipment. As opposed to air cooling, water is used as the heat conductor. Water cooling is commonly used for cooling automobile internal combustion engines and large industrial facilities such as steam electric power plants, hydroelectric generators, petroleum refineries and chemical plants. Other uses include cooling the barrels of machine guns, cooling of lubricant oil in pumps; for cooling purposes in heat exchangers; cooling products from tanks or columns, and recently, cooling of various major components inside high-end personal computers. The main mechanism for water cooling is convective heat transfer.

NOMENCLATURE

Cooling water is the water removing heat from a machine or system. Cooling water may be recycled through a recirculating system or used in a single pass once-through cooling (OTC) system. Recirculating systems may be open if they rely upon cooling towers or cooling ponds to remove heat or closed if heat removal is accomplished with negligible evaporative loss of cooling water. A heat exchanger or condenser may separate non-contact cooling water from a fluid being cooled, or contact cooling water may directly impinge on items like saw blades where phase difference allows easy separation. Environmental regulations emphasize the reduced concentrations of waste products in non-contact cooling water.

ADVANTAGES

Water is inexpensive and non-toxic. The advantages of using water cooling over air cooling include water's higher specific heat capacity, density, and thermal conductivity. This allows water to transmit heat over greater distances with much less volumetric flow and reduced temperature difference.

For cooling CPU cores in computing equipment, the primary advantage of water cooling is that it is capable of transporting heat away from the source to a secondary cooling surface to allow for large, more optimally designed radiators rather than small, relatively inefficient fins mounted directly on the heat source.

The water jacket around an engine is also very effective at deadening mechanical noises, which makes the engine quieter. However, the primary disadvantage is that it costs significantly more than an air-cooled engine system.

Disadvantages

Water accelerates corrosion of metal parts and is a favorable medium for biological growth. Dissolved minerals in natural water supplies are concentrated by evaporation to leave deposits called scale. Cooling water often requires addition of chemicals to minimize corrosion and insulating deposits of scale and biofouling.In water cooling systems for electronic devices, the use of deionized water is required, which must be carefully controlled in order to avoid contamination, which would cause a decrease in resistance of the water and subsequently increase risk of short circuits.

OPEN METHOD

An antique gasoline engine with an evaporative cooler and mesh screen to improve evaporation. Water is pumped up to the top and flows down the screen to the tank.

An open water cooling system makes use of evaporative cooling, lowering the temperature of the remaining (unevaporated) water. This method was common in early internal combustion engines, until scale buildup was observed from dissolved salts and minerals in the water. Modern open cooling systems continuously waste a fraction of recirculating water as blowdown to remove dissolved solids at concentrations low enough to prevent scale formation. Some open systems use inexpensive tap water, but this requires higher blowdown rates than deionized or distilled water. Purified water systems still require blowdown to remove accumulation of byproducts of chemical treatment to prevent corrosion and biofouling.

Automotive Usage

Pressurization

Modern automotive cooling systems are slightly pressurized, often to 15 psi (103 kPa). This raises the boiling-point of the coolant and reduces evaporation.

Antifreeze

The use of water cooling carries the risk of damage from freezing. Automotive and many other engine cooling applications require the use of a water and antifreeze mixture to lower the freezing point to a temperature unlikely to be experienced. Antifreeze also inhibits corrosion from dissimilar metals and can increase the boiling point, allowing a wider range of water cooling temperatures. Its distinctive odor also alerts operators to cooling system leaks and problems that would go unnoticed in a water-only cooling system. The heated coolant mixture can also be used to warm the air conditioning system inside the car by means of the heater core.

Other additives

Other less common chemical additives are products to reduce surface tension. These additives are meant to increase the efficiency of automotive cooling systems. Such products are used to enhance the cooling of underperforming or undersized cooling systems or in racing where the weight of a larger cooling system could be a disadvantage.

Power Electronics And Transmitters

Since approximately 1930 it is common to use water cooling for tubes of powerful transmitters. As these devices uses high operation voltages (around 10 kV), the use of deionized water is required and it has to be carefully controlled. Modern solid-state transmitters can be built so, that even high power transmitters do not require water cooling. Water cooling is however also sometimes used for thyristors of HVDC valves, for which also the use of deionized water is required.

Computer Usage

Cooling hot computer components with various fluids has been in use since at least as far back as the development of Cray-2 in 1982, using Fluorinert. Through the 1990s, water cooling for home PCs slowly gained recognition amongst enthusiasts, but it started to become noticeably more prevalent after the introduction of AMD's hot-running Athlon processor in mid-2000. As of 2011, there are several manufacturers of water cooling components and kits, and some custom computer retailers include various setups of water cooling for their high-performance systems.

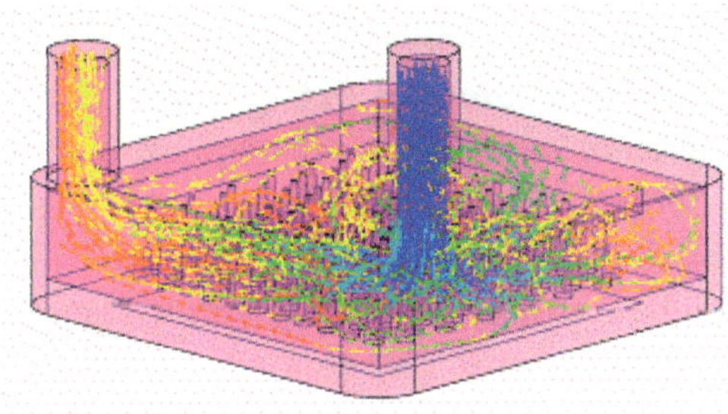

This 40 mm by 40 mm by 10 mm impingement-type liquid-cooled cold plate (heat sink) animation was created using high-resolution CFD analysis, and shows temperature contoured flow trajectories, predicted using a CFD analysis package, courtesy of NCI.

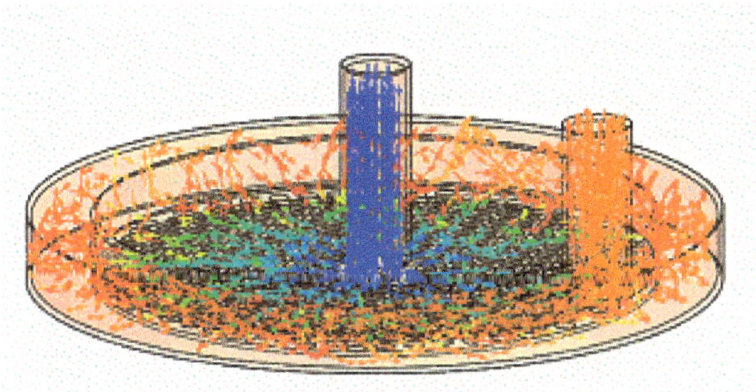

This 60mm diameter by 10mm high impingement-type water-cooled copper cold plate (heat sink) animation was created using high-resolution CFD analysis, and shows temperature contoured flow trajectories, predicted using a CFD analysis package, courtesy of NCI.

Water cooling can be used to cool many computer components, but especially the CPU. Water cooling usually uses a CPU water block, a water pump, and a water-to-air heat exchanger. By transferring device heat to a separate heat exchanger which can variously be made large and use larger, lower-speed fans, Water cooling can allow quieter operation, improved processor speeds (overclocking), or a balance of both. Less commonly, GPUs, Northbridges, Southbridges, hard disk drives, memory, voltage regulator modules (VRMs), and even power supplies can be water-cooled.

Radiator size may vary: from 40mm dual fan (80mm) to 140 quad fan (560mm) and thickness from 30mm to 80mm

Water coolers for desktop computers were, until the end of the 1990s, homemade. They were made from car radiators (or more commonly, a car's heater core), aquarium pumps and home-made water blocks, laboratory-grade PVC and silicone tubing and various reservoirs (homemade using plastic bottles, or constructed using cylindrical acrylic or sheets of acrylic, usually clear) and or a T-Line. More recently a growing number of companies are manufacturing water-cooling components compact enough to fit inside a computer case. This, and the trend to CPUs of higher power dissipation, has greatly increased the popularity of water cooling, although only a very small minority of computers are water-cooled.

Dedicated overclockers occasionally use vapor-compression refrigeration or thermoelectric coolers in place of more common standard heat exchangers. Water cooling systems in which water is cooled directly by the evaporator coil of a phase change system are able to chill the circulating coolant below the ambient air temperature (impossible with a standard heat exchanger) and, as a result, generally provide superior cooling of the computer's heat-generating components. The downside of phase-change or thermoelectric cooling is that it uses much more electricity, and antifreeze must be added due to the low temperature. Additionally, insulation, usually in the form of lagging around water pipes and neoprene pads around the components to be cooled, must be used in order to prevent damage caused by condensation of water vapour from the air on chilled surfaces. Common places from which to borrow the required phase transition systems are a household dehumidifier or air conditioner.

An alternative cooling system, which enables components to be cooled below the ambient temperature, but which obviates the requirement for antifreeze and lagged pipes, is to place a thermoelectric device (commonly referred to as a 'Peltier junction' or 'pelt' after Jean Peltier, who documented the effect) between the heat-generating component and the water block. Because the only sub-ambient temperature zone now is at the interface with the heat-generating component itself, insulation is required only in that localized area. The disadvantage of such a system is a higher power dissipation.

To avoid damage from condensation around the Peltier junction, a proper installation requires it to be "potted" with silicone epoxy. The epoxy is applied around the edges of the device, preventing air from entering or leaving the interior.

Apple's Power Mac G5 was the first mainstream desktop computer to have water cooling as standard. Dell followed suit by shipping their XPS computers with liquid cooling, using thermoelectric cooling to help cool the liquid. Currently, Dell's only computers to offer liquid cooling are their Alienware desktops.

Liquid Cooling Maintenance

Liquid cooling techniques are increasingly being used for the thermal management of electronic components. This type of cooling is a solution to ensure the optimisation of energy efficiency while simultaneously minimising noise and space requirements. Especially useful in supercomputers or Data Centers as maintenance of the racks is quick and easy. After disassembly of the rack, advanced technology quick release couplings eliminate spillage for the safety of operators and protects the integrity of fluids (no impurities in the circuits). These couplings are also capable of being locked (Panel mounted?) to allow blind connection in difficult to access areas. It is important in electronics technology to analyse the connection systems to ensure:

- Non-spill sealing (clean break, flush face couplings)
- Compact and lightweight (materials in special aluminum alloys)
- Operator safety (disconnection without spillage)

CoolIT Rack DCLC AHx Liquid Cooling Solution

- Quick-release couplings sized for optimized flow
- Connection guiding system and compensation of misalignment during connection on rack systems
- Excellent resistance to vibration and corrosion
- Designed to withstand a large number of connections even on refrigerant circuits under residual pressure

A Marley mechanical induced draft cooling tower

Industrial Usage

Industrial cooling towers may use river water, coastal water (seawater), or well water as their source of fresh cooling water. The large mechanical induced-draft or forced-draft cooling towers in industrial plants continuously circulate cooling water through heat exchangers and other equipment where the water absorbs heat. That heat is then rejected to the atmosphere by the partial evaporation of the water in cooling towers where upflowing air is contacted with the circulating downflow of water. The loss of evaporated water into the air exhausted to the atmosphere is replaced by "make-up" fresh river water or fresh cooling water. Since the evaporation of pure water is replaced by make-up water containing carbonates and other dissolved salts, a portion of the circulating water is also continuously discarded as "blowdown" water to prevent the excessive build-up of salts in the circulating water.

The Indian Point Energy Center. Over a billion fish eggs and larvae are killed in its cooling system each year.

Cooling water intake of a nuclear power plant

On very large rivers, but more often at coastal and estuarine sites, "direct cooled" systems are often used, instead. These industrial plants do not use cooling towers and the atmosphere as a heat sink, but put the waste heat to the river or coastal water instead. These OTC systems thus rely upon a good supply of river water or seawater for their cooling needs. Many facilities, particularly electric power plants, use millions of gallons of water per day for cooling. Such facilities are built with intake structures designed to pump in large volumes of water at a high rate of flow. These structures tend to also pull in large numbers of fish and other aquatic organisms, which are killed or injured on the intake screens.

The warmed water is returned directly to the aquatic environment, often at temperatures significantly (to aquatic life) above the ambient receiving water. Thermal pollution of rivers, estuaries and coastal waters is a consideration when siting such plants.

High-grade industrial water (produced by reverse osmosis) and potable water are sometimes used in industrial plants requiring high-purity cooling water.

Some nuclear reactors use heavy water as cooling. Heavy water is employed in nuclear reactors because it is a weaker neutron absorber. This allows for the use of less enriched fuel. For the main cooling system, normal water is preferably employed through the use of a heat exchanger, as heavy water is much more expensive. Reactors that use other materials for moderation (graphite) may also use normal water for cooling.

Marine Vessel Use

Water is an ideal cooling medium for vessels as they are constantly surrounded by water that generally remains at a low temperature throughout the year. This does however pose new challenges as cooling systems operating with sea water need to be manufactured from materials that are suitable for the environment. A heat exchanger for example will need to be manufactured from materials such as Cupronickel, Bronze or Titanium to protect it from erosion or corrosion. The velocity will also need to be far more restricted compared with a fresh water cooling system. If the velocity is too low; then sand and other sediments can block the tubes. If the velocity is too high then the tubes can be eroded by the sediments in the water.

ENVIRONMENTAL CONSIDERATIONS

Water is a favorable environment for many life forms. Water cooling may alter natural water environments and create new environments. Flow characteristics of recirculating cooling water systems encourage colonization by sessile organisms to use the circulating supply of food, oxygen and nutrients. Volumes of water lost during evaporative cooling may decrease natural habitat for aquatic organisms. Water temperature increases modify aquatic habitat by increasing biochemical reaction rates and decreasing oxygen saturation capacity of the habitat. Temperature increases initially favor a population shift from those requiring the high-oxygen concentration of cold water to those enjoying advantages of increased metabolic rates in warm water. Temperatures may become high enough to support thermophilic populations.

Biofouling of heat exchange surfaces can reduce heat transfer rates of the cooling system; and biofouling of cooling towers can alter flow distribution to reduce evaporative cooling rates. Biofouling may also create differential oxygen concentrations increasing corrosion rates. OTC and open recirculating systems are most susceptible to biofouling. Biofouling may be inhibited by temporary habitat modifications. Temperature differences may discourage establishment of thermophilic populations in intermittently operated facilities; and intentional short term temperature spikes may periodically kill less tolerant populations. Biocides have been commonly used to control biofouling where sustained facility operation is required.

Large OTC flow rates may immobilize slow-swimming organisms including fish and shrimp on screens protecting the small bore tubes of the heat exchangers from blockage. High temperatures or pump turbulence and shear may kill or disable smaller organisms passing the screens entrained with the cooling water. In the U.S., cooling water intake structures kill billions of fish and other organisms each year. More agile aquatic predators consume organisms impinged on the screens; and warm water predators and scavengers colonize the cooling water discharge to feed on entrained organisms.

Manufactured metals tend to revert to ores via electrochemical reactions of corrosion. Water can accelerate corrosion as both an electrical conductor and solvent for metal ions and oxygen. Corrosion reactions proceed more rapidly as temperature increases. Preservation of machinery in the presence of hot water has been improved by addition of chemicals including zinc, chromates and phosphates. The first two have toxicity concerns; and the last has been associated with eutrophication. Residual concentrations of biocides and corrosion inhibitors are of potential concern for OTC and blowdown from open recirculating systems.With the exception of machines with short design life, closed recirculating systems require periodic cooling water treatment or replacement raising similar concern about ultimate disposal of cooling water containing chemicals used with environmental safety assumptions of a closed system.

Industrial Cooling Water Regulations

The U.S. Clean Water Act requires the Environmental Protection Agency (EPA) to issue regulations on industrial cooling water intake structures. EPA issued final regulations for new facilities in 2001 (amended 2003), and for existing facilities in 2014.\

COOLING WATER CHEMISTRY

Water contains varying amounts of impurities from contact with the atmosphere, soil, and containers. Cooling water treatments add other chemicals attempting to maintain satisfactory heat exchange.

Solids

Total dissolved solids or TDS (sometimes called filtrable residue) is measured as the mass of residue remaining when a measured volume of filtered water is evaporated. Salinity measures water density or conductivity changes caused by dissolved materials.

Probability of scale formation increases with increasing total dissolved solids. Solids commonly associated with scale formation are calcium and magnesium carbonate and sulfate. Corrosion rates initially increase with salinity in response to increasing electrical conductivity, but then decrease after reaching a peak as higher levels of salinity decrease dissolved oxygen levels.

Hydrogen

Water ionizes into hydronium (H_3O^+) cations and hydroxide (OH^-) anions. The concentration of ionized hydrogen (as protonated water) is expressed as pH. Low pH values increase rate of corrosion while high pH values encourage scale formation. Amphoterism is uncommon among metals used in water cooling systems, but aluminum corrosion rates increase with pH values above 9. Galvanic corrosion may be severe in water systems with copper and aluminum components. Acid may be added to cooling water systems to prevent scale formation if the pH decrease will offset increased salinity and dissolved solids.

Phosphorus and chromium

Concentrations of polyphosphates or phosphonates with zinc and chromates or similar compounds have been maintained in cooling systems to keep heat exchange surfaces clean so a film of gamma iron oxide and zinc phosphate may inhibit corrosion by passivating anodic and cathodic reaction points. These increase salinity and total dissolved solids, and phosphorus compounds may provide the limiting essential nutrient for algal growth contributing to biofouling of the cooling system or to eutrophication of natural aquatic environments receiving blowdown or OTC water. Chromates reduce biofouling in addition to effective corrosion inhibition, but residual toxicity in blowdown or OTC water has encouraged reduced chromate concentrations and use of less flexible corrosion inhibitors. Blowdown may also contain chromium leached from cooling towers constructed of wood preserved with chromated copper arsenate.

Oxygen

Some groundwater contains very little oxygen when pumped from wells, but most natural water supplies include dissolved oxygen. Corrosion increases with increasing oxygen concentrations. Dissolved oxygen approaches saturation levels in cooling towers. Dissolved oxygen is desirable in blowdown or OTC water being returned to natural aquatic environments.

Biocides

Chlorine may be added in the form of hypochlorite to decrease biofouling, but is later reduced to chloride to minimize toxicity of blowdown or OTC water returned to natural aquatic environments. Hypochlorite is increasingly destructive to wooden cooling towers as pH increases. Chlorinated phenols have been used as biocides or leached from preserved wood in cooling towers. Both hypochlorite and pentachlorophenol have reduced effectiveness at pH values greater than 8. Non-oxidizing biocides may be more difficult to detoxify prior to release of blowdown or OTC water to natural aquatic environments.

REFERENCES

1. U.S. Environmental Protection Agency (EPA). (1997). Profile of the Fossil Fuel Electric Power Generation Industry (Report). Washington, D.C. Document No. EPA/310-R-97-007. p. 79.
2. http://www.che.ufl.edu/unit-ops-lab/experiments/CT/2%20CT-Theory.pdf
3. Kemmer (1979), pp. 38-1, 38-4, 38-7 & 38-8.
4. King (1995), pp. 143, 439.
5. Betz, pp. 183–184.
6. Betz, p. 192.
7. Nice, Karim. "How Car Cooling Systems Work". HowStuffWorks. HowStuffWorks, Inc. Retrieved 20 August 2012.
8. http://koolance.com/1300-1700w-liquid-cooled-power-supply
9. «Alienware Desktops". Dell. Retrieved 2009-11-05.
10. Beychok, Milton R. (1967). Aqueous Wastes from Petroleum and Petrochemical Plants (1st ed.). John Wiley and Sons. LCCN 67019834. (See Chapter 2 for material balance relationships in a cooling tower)
11. McGeehan, Patrick (2015-05-12). "Fire Prompts Renewed Calls to Close the Indian Point Nuclear Plant". New York Times.
12. EPA (2010). "Partial List of Facilities Subject to Clean Water Act 316(b)."
13. EPA (2014). "Cooling Water Intakes."
14. Thermex "Heat Exchanger FAQ Page" 2016-12-12.
15. Reid (1961), pp. 267–268.
16. Betz, p. 202.
17. EPA (2004). "Economic and Benefits Analysis for the Final Section 316(b) Phase II Existing Facilities Rule." Chapter A2. Document No. EPA-821-R-04-005.
18. EPA (May 2014). "Final Regulations to Establish Requirements for Cooling Water Intake Structures at Existing Facilities." Fact sheet. Document no. EPA-821-F-14-001.
19. Goldman & Horne (1983), pp. 153, 160.
20. Clean Water Act, Section 316(b), 33 U.S.C. § 1316.
21. EPA. Cooling Water Intake Structures. Final rule: 2001-12-18, 66 FR 65255. Amended: 2003-06-19, 68 FR 36749.
22. EPA. "National Pollutant Discharge Elimination System—Final Regulations To Establish Requirements for Cooling Water Intake Structures at Existing Facilities and Amend Requirements at Phase I Facilities" Final rule. Federal Register, 79 FR 48300. 2014-08-15.

Index